P9-DWZ-567

The Biosphere

The Biosphere

Ian K. Bradbury

Belhaven Press
(a division of Pinter Publishers)
London and New York

First published in Great Britain in 1991 by
Belhaven Press (a division of Pinter Publishers),
25 Floral Street, London WC2E 9DS

British Library Cataloguing in Publication Data

A CIP catalogue record for this book is available from the
British Library

ISBN 1 85293 037 3 (hb)
     1 85293 038 1 (pb)

For enquiries in North America please contact
PO Box 197, Irvington, New York 10533

Library of Congress Cataloging in Publication Data

Bradbury, Ian K., 1944–
  The biosphere/Ian K. Bradbury.
    p.   cm.
  Includes bibliographical references and index.
  ISBN 1-85293-037-3. – ISBN 1-85293-038-1 (pbk.)
  1. Ecology. 2. Biosphere. 3. Paleontology. I. Title.
QH541.B697   1991
574.5–dc20
                                                                    90-24076
                                                                         CIP

Filmset by Mayhew Typesetting, Bristol, England
Printed and bound by Biddles Ltd of Guildford and Kings Lynn

Contents

Preface

An understanding of how our environment works and how it responds to human activities requires some knowledge of living organisms and ecological systems. Yet many students enter higher education to pursue environmentally-centred courses such as geography with little or no formal training in the life sciences. This book is written primarily for such students, but also for anyone without a biological training who is seeking a basic understanding of the biosphere. In England and Wales (but not in Scotland) we retain a narrow – and increasingly indefensible – curriculum for 16–18 year olds whose minds inevitably become compartmentalized quite early. An unfortunate result is that subjects not studied in the last two years of secondary education are frequently regarded as inaccessible which applies particularly to the sciences. This book attempts to introduce the basic functional features of the biosphere so as to make one component of our natural environment more accessible to the non-specialist. No assumptions are made about biological – or chemical – knowledge beyond what might be reasonably regarded as 'everyday', and there are certainly no prerequisites in terms of courses successfully completed.

In practically all institutions offering degree level work in geography, students are required to undertake some work in the physical environment. Geographers traditionally accept they must know something about the composition of the Earth and the processes that shape its surface, and frequently acknowledge that hydrology and atmospheric phenomena are also part of the deal. With the 'living world' there appears to be less certainty; a tendency exists to regard that part, labelled 'biology', as a separate, and somewhat alien, discipline. However, geographers are concerned with an impressively wide range of subject matter requiring at least some knowledge of organisms and ecology. In addition to the Earth's natural systems such topics include environmental history, land form development, climatic change, environmental impact assessment, agricultural practices, pollution and resource conservation. Unfortunately, basic biological principles relevant to such topics are all too frequently neglected by geography students, and also their teachers. Consequently the level of understanding is sometimes rather more superficial than is desirable. It is hoped that this review of some fundamental concepts, with its emphasis on functional aspects, will provide a foundation for the study of these and other such issues which are central to the concerns of geographers and others engaged in environmental studies.

In deciding what to include in this text I was guided chiefly by what I felt I would like my own students to know about, and also by the sorts of questions they frequently ask. Of course, opinions will differ concerning the choice of subjects, the relative amounts of time spent on each and the arrangement of the chosen material. As far as structure is concerned, I have tried not to introduce a topic without first providing some background necessary for its understanding. Due consideration has been given to topics such as aquatic ecosystems and decomposition, frequently given short shrift in 'biogeography' texts, but which are central to an understanding of the biosphere. On the other hand I have made little reference to regional level distributions of the biota – i.e. biogeography *sensu stricto* – because this subject is well treated elsewhere for a similar audience.

One of the major problems in producing an elementary treatment of life processes is that there always appear to be exceptions to generalized statements of principle. Accordingly, a great many more caveats might have been justified than actually appear. However, my aim was to establish a set of workable frameworks without obscuring the points of central concern. There was a similar problem in deciding just how much vocabulary – much of it likely to be new – to include, and again compromises were necessary.

Many environmental texts conclude with a 'human impact' chapter. I have avoided this practice, partly because such chapters are usually too short to be satisfying, but chiefly because human activities tend to affect the rate and degree of 'natural' processes rather than differing to them in kind. To reinforce this principle, some major environmental problems are mentioned where relevant processes are being discussed.

During the preparation of chapter drafts, I relied heavily on the critical comments of others; most were asked because of their particular expertise, but some provided a non-specialist view. For their valuable time, support and guidance I am extremely grateful to Christopher Beadle, Mary Benbow, Andrew Charlesworth, Julian Collins, Stephen Cuttle, John Gode, John Goode, Angus Gunn, Adrian Harvey, Ann Henderson-Sellers, Peter James, Meriel Jones, Cedric Milner, Christopher Paul, Philip Putwain, Marjorie Sullivan, David Thompson, Heather Paterson, Judith Quinn, Barbara Rouse, Kent Walton and Bernard Wood. Of course, responsibility for the finished product lies with the author.

Special thanks are due to Paul Smith, who drew all the diagrams, and to Ian Qualtrough and Suzanne Yee who were responsible for photographic work. Appreciation is expressed to the Trustees of the National Museums and Galleries on Merseyside for granting access to the collections of the Liverpool Museum and to Eric Greenwood, Keeper of the Liverpool Museum, for facilitating the reproduction of material. Philip W. Phillips, Curator of Palaeontology at the Liverpool Museum, is warmly thanked for much invaluable guidance and advice. Thanks are also due to Iain Stevenson of Belhaven Press for his encouragement and support and to Vanessa Harwood who competently oversaw the final stages of production.

Acknowledgements

The author gratefully acknowledges permissions to use the following data and illustrated material. Table 8.1 From Humphreys, W. (1979) *Journal of Animal Ecology* 48: 427. Fig. 9.1 From Watts, C.H.F. (1968) *Journal of Animal Ecology* 37: 25–41. Redrawn with permission. Blackwell Scientific Publications Limited. Fig. 9.2 From Swift, M.J. *et al* (1981) *Journal of Ecology* 69: 981–95. Redrawn with permission. Blackwell Scientific Publications Limited. Fig. 9.3 From Varley, G.C. (1970) in: Watson, A. (Ed) *Animal Populations in Relation to their Food Resources*. Redrawn with permission. Blackwell Scientific Publications Limited. Fig. 10.5 Data supplied by Peter James and Andrew Wharfe. Fig. 10.6 From Bormann, F.H and Likens, G.E. (1979) *Pattern and Process in a Deforested Ecosystem*. Redrawn with permission. Springer-Verlag. Fig. 10.7 From Centre for Agricultural Strategy (1978) *Phosphorus: A Resource for UK Agriculture*. University of Reading. Redrawn with permission Professor Spedding. Fig. 10.9 From Barnes, R.K.S and Mann, K.H. (1980) *Fundamentals of Aquatic Ecosystems*. Copied with permission. Blackwell Scientific Publications Limited. Fig 10.10 Redrawn from data in Boden, T.A *et al* (1990) *Trends '90: a compendium of data on global change*. The carbon dioxide information analysis centre, Oak Ridge, Tennessee. Based on analysis of measurements made at Mauna Loa, Hawaii; principal invesigators C.D. Keeling and T.P Whorf. Figs. 12.1, 13.6, 13.7, 13.9, 14.1, 14.2, 14.4, 14.5, 14.6, 14.8, 14.10, 14.11. With kind permission of the Trustees of the National Museums and Galleries on Merseyside. Fig 13.2 From Ridley, B.K. 1979 *The Physical Environment*. Redrawn with permission Ellis Horwood Limited. Fig 13.4 Redrawn with permission from Jones, G. *et al. Dictionary of Environmental Sciences* (1990) Harper Collins Publishers. Fig. 13.10 From Cox, C.B. *et al* (1976) *Biogeography – An Ecological and Evolutionary Approach* 2nd edn. Copied with permission. Blackwell Scientific Publications Limited. Fig 14.3 Photo Ian Qualtrough.

Part 1 *Unifying principles*

In this set of chapters we concentrate on principles and processes that are important for an understanding of life on Earth. We shall see that some basic principles apply universally to living organisms, regardless of their size, shape, and ecological role. We shall also see that an understanding of function in the biosphere demands some familiarity with the basic metabolic processes that go on within individual organisms. The overall aim here is to establish a framework for thinking about living processes, and to introduce a basic working vocabulary.

First, we shall identify important levels of organization to order our thinking. In the chapter that follows we consider the chemical basis for life. Then, in successive chapters, we introduce some fundamentals of growth and reproduction, the idea of evolution, and energy and life. Having emphasized the unifying characteristics of life we conclude this part of the book by considering the variety of organisms on Earth and how they are classified and named.

1 *Levels of organization*

The term 'biosphere' is variously defined, but it is used principally in two ways. Perhaps most commonly, the biosphere encompasses all the zones of the Earth in which life is present. However the term is sometimes used to refer only to the sum total of the organisms on the planet. Used in the latter sense, the term raises certain questions; what about the remains of organisms, for example; should these be considered part of the biosphere along with the living biota; should anything of organic origin be considered part of the biosphere? If so, does the biosphere include the coal measures that formed more than 300 million years ago?

Although there may be problems in agreeing on a close definition of 'biosphere', that is not important. What is important is that there is widespread acceptance of the set of *ideas* conveyed by the term. The idea of living organisms as functional entities, processing energy and matter, reproducing, interacting with each other and with the physical environment; and the idea that the entire biota, together with its physical environment, can be regarded as a single system.

As a point of departure we identify some logical and convenient levels of organization to order our thoughts about life on our planet. This will allow us to establish some reference points and also to introduce certain key concepts together with some relevant terminology.

Organisms

Perhaps the most convenient level of organization with which to begin is the individual organism. Most familiar organisms appear to be discrete entities; an oak tree and a mouse, for example, are apparently distinct morphologically and functionally. For most types of organism the concept of the individual holds true, but it is worth pointing out that some organisms form colonies in which it is not easy to distinguish functionally-discrete individuals. Even apparently distinct individuals such as trees may form connections between their root systems. To make things more difficult, certain organisms, ants for example, display such a high degree of social organization, even though they are not tangibly connected, that the notion of the 'individual' might be questioned. However, these qualifications need not concern us unduly, and the idea of the individual as representing a useful level of organization for most practical purposes will serve for now.

The cellular structure of organisms

All organisms are made up of *cells*. The cell is a fundamentally important unit of biological organization because it is within cells that the vital processes of life occur. We shall make frequent reference to cells and cellular processes, particularly in the first part of the book, because understanding at the cellular level of organization will enhance appreciation of the functional aspects of ecological systems and the whole biosphere.

Some types of organism – a bacterium and an amoeba for example – comprise just one cell, and are appropriately referred to as *unicellular* or *single-celled* organisms. All other organisms are *multicellular*. The number of cells comprising a large animal, such as an individual of our own species, is difficult to comprehend because it runs into hundreds of billions. The processes of cell multiplication, which are central to growth and reproduction receive some attention in Chapter 3.

Cells are essentially microscopic entities, meaning they cannot be discerned with the naked eye. It is true that certain types of cell from some organisms are visible without magnification but these are exceptions. Because of their very small size, cell dimensions are usually expressed in micrometres (μm); one micrometre is just one millionth of a metre. Bacterial cells are typically between 0.5 and 1.0μm in diameter whereas a human egg cell has a diameter of about 100μm. As we shall see later, all cells from a multicellular organism share certain features in common, even though they may appear different and perform specialized functional roles. Aggregations of cells of broadly similar appearance and function are known as *tissues*. The term *organ* is used to refer to an organism's parts which are well defined morphologically and functionally; the heart and liver for example.

At this stage it will be useful to note that all cells are described as either *prokaryotic* or *eukaryotic*. Furthermore, all organisms are either prokaryotic or eukaryotic depending on their cell type. Single-celled organisms may either be of prokaryotic or eukaryotic type, but all true multicellular organisms are eukaryotic. These two cell types are fundamentally different in organization. As we progress through the early chapters, the main points of distinction between prokaryotic and eukaryotic cells (and organisms) will be pointed out. For now, we may note that eukaryotic cells are the larger of the two cell types and that they have a greater degree of structural organization. In eukaryotic cells, many of the essential processes take place within *organelles*, a general term used to refer to subcellular, or intracellular, structures that are bounded by membranes. The *nucleus* is probably the most widely known of these, but there are several others. In contrast, prokaryotic cells do not have true organelles. In general, however, similar functional processes go on in a prokaryotic cell as in a eukaryotic cell; it is just that in a prokaryotic cell these processes do not take place within well-defined subcellular structures but in the cellular *cytoplasm*.

Cells from different kinds of organism have characteristic features.

Cells of plants, fungi and bacteria, for example, are surrounded by a cell wall, while much of the interior of plant cells is typically occupied by a fluid-filled cavity known as a *vacuole*. Consideration of subcellular levels of organization would soon take us into the realms of chemistry, but we leave this important topic until Chapter 2 and now look outward from the individual organism.

Populations

A group of individual organisms of the same type living within a prescribed area is known as a *population*. We may refer to the population of tigers in a national park in India, the population of oak trees in a forest, or the human population of a named country. The population is a major focus of interest for ecologists and evolutionary biologists. Much attention is given to determining population size (abundance), to how population size changes over time, to the factors that determine abundance and to how population characteristics change with time. Population size is a function of the rate at which new individuals are 'born' (*natality*), the rate at which individuals die (*mortality*) and also the rate at which individuals move in or out of the area of interest. So understanding what factors influence these rates is a key area of ecological research, and it is one with major implications for conservation, natural resource management and the control of pest populations.

Communities and ecosystems

Populations do not naturally occur on their own; rather, they coexist with other populations to form ecological *communities*. Unlike cells, which demonstrably exist as well defined units, the community is a rather abstract level of organization. What is meant here is that a community does not exist as a discrete entity, as does an individual cell with its clearly defined boundary. Any such demarcation of a collection of plants and animals would be quite arbitrary, although it may well be necessary for practical purposes. In fact, the biosphere is best considered as being continuously variable in space in terms of environment and biotic details, even though superficial examination may suggest a high degree of uniformity over large areas.

Despite the fact that communities are not naturally demarcated, the concept of the community as an assemblage of interacting organisms of different species is still valid and, rather like the term 'biosphere', it is the ideas conveyed that are really important. The term community is widely employed and the organization of communities is another major focus of ecological study. Moreover, as particular types of environment in different areas often support forms of life which have a degree of morphological similarity and functional equivalence, we can refer informally to 'community types'; thus we may talk about a sand dune community, a heathland community, an insect community in a corn-field, a pond community and so on.

The remarks just made about the community apply also to *ecosystem*, a term used to imply the functional nature of communities and the various interactions that occur between a community of organisms and its physical environment. So expressions such as marine ecosystem, fresh-water ecosystem, tropical forest ecosystem and agricultural ecosystem are commonly employed, implying that each type has certain distinguishing features. Understanding the organization and function of these various types of ecosystem requires observation and experimental work, either in what seem to be good representatives of the type, or in artificial situa-tions which we regard as relevant to the natural world. But as is the case with communities, we should not regard the Earth's surface as compris-ing a set of ecosystems with clearly defined boundaries; the real world is much less tidy. The term *biome* is used to refer to regional-level land biotas with similar structural and functional features; for example, the tundra biome and hot desert biome.

2 *The chemical basis for life*

In this chapter we focus on chemical aspects of the biosphere. The objective here is to provide for the reader with no scientific background some basic material so that what follows later in the book will be more accessible and better appreciated. For many readers much of this material will seem very elementary; others may think it a bit dull, and possibly daunting, to be faced with some chemistry at the outset, but the fact is we simply cannot avoid some chemical terminology when discussing the biosphere. After all, living processes are basically chemical processes. The aim here is not to frighten, but rather to facilitate greater understanding of the various topics we encounter in this book. We shall be working at a very descriptive level, so any preconceived notions about difficulties of understanding anything remotely scientific should be immediately discarded.

Atoms and elements

All substances, whether in gaseous, liquid or solid form, are composed of extremely tiny entities called *atoms*. Atoms are made up of even smaller particles, the most important of which for our purposes are *protons*, *neutrons* and *electrons*. The protons and neutrons essentially comprise the atomic nucleus, around which the electrons are in motion. A proton carries a unit positive charge; an electron carries a unit negative charge. An atom with the same number of protons and electrons is therefore electrically neutral.

An atom is characterized, and distinguished, by the number of protons in its nucleus. The number of protons in naturally occurring atomic nuclei, a value known as the *atomic number*, ranges between one and a little over a hundred. An *element* is a substance composed of atoms with the same number of protons in their nuclei. Atoms with different numbers of protons in their nuclei are of different elements, and they differ in their chemical properties. Most of the elements occur naturally but a few can only be produced artificially by bombarding certain types of atomic nuclei with protons. Each chemical element is represented by a shorthand symbol consisting of one or two letters. For many elements, e.g. carbon (C), nitrogen (N) and phosphorus (P), this is the first letter of the English word, but for many other elements, e.g. lead (Pb), potassium (K) and mercury (Hg), the symbol is derived from another language, usually Latin.

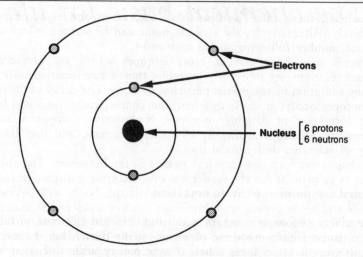

Fig. 2.1 Schematic representation of an atom of carbon-12 (^{12}C), the most abundant
isotope of carbon. In reality the nucleus occupies only one ten-thousandth the
volume of the atom. Atomic number (number of protons) is 6; mass number
(number of protons plus number of neutrons) is 12. The electrons are associated
with two shells. Living organisms are composed principally of carbon-based
substances.

A single atom of carbon is represented schematically in Fig. 2.1. The
picture of electrons moving around the nucleus in circular orbits or shells
(known as the planetary model) is a convenient, but rather simplistic
representation of reality. (In fact, electrons are associated with volumes,
known as *orbitals*.) The atoms of each element have a characteristic
number of shells, each with a characteristic number of electrons. Note
that the carbon atom in Fig. 2.1 has two electron shells, one with two
electrons, the other with four electrons. The importance of atomic structure will soon become clear.

Isotopes

The *mass number* of an atom is the sum of the number of protons and
the number of neutrons. Now although the number of protons is the
same for all atoms of an individual element, the number of neutrons may
differ. So atoms of the same element do not necessarily have the same
mass number. Atoms of the same element that differ in the number of
neutrons are called *isotopes* of that element. Nearly all the naturally
occurring elements exist in more than one isotopic form, although
frequently just one is overwhelmingly predominant. To show which
isotope is being referred to, it is conventional to use the appropriate mass
number as a superscript preceding the symbol for the element. Thus the
three naturally occurring isotopes of carbon are represented as ^{12}C, ^{13}C
and ^{14}C. As carbon has an atomic number of six (i.e. six protons in

the nucleus), these three isotopes have six, seven and eight neutrons respectively. Alternatively, the element name can be written in full with the mass number following, as in carbon-14.

Elements may have one or more isotopes which are *radioactive*. Radioactive atoms are inherently unstable; they *decay* spontaneously, by releasing radiation or subatomic particles, or both. The decay of different radioisotopes occurs at vastly different, but characteristic, rates and leads to the formation of daughter isotopes. Radioactive isotopes of many elements are produced artificially in nuclear reactors, and they have a variety of scientific and clinical uses.

Elemental composition of living organisms

Of the ninety-two or so naturally occurring chemical elements on Earth, living organisms have made use of about twenty-five. Most of these are found in every living cell, regardless of type and regardless of organismic origin. Chemical elements that are essential for life processes are called *nutrients*.

Carbon is the principal structural element for all living organisms, typically comprising around half of the dry weight of an animal or plant. Two other elements, oxygen and hydrogen, are closely associated with carbon; oxygen makes up around 45 per cent and hydrogen a little over 5 per cent of the dry weight. (If we included water, which makes up a significant proportion of the *total* weight of an organism, obviously the proportional contribution of hydrogen and oxygen would be much greater.) Carbon, hydrogen and oxygen, are often treated separately from the other essential elements, partly because of their structural contribution to living tissue but also because their mode of acquisition by some groups of organisms is rather different from the other elements. In addition to carbon, hydrogen and oxygen most of the following listed elements are present in all living cells.

	F	fluorine	
	Na	sodium	
	Mg	magnesium	
N nitrogen	Cl	chlorine	Si silicon
P phosphorus	K	potassium	V vanadium
S sulphur	Ca	calcium	Sn tin
	Mn	manganese	I iodine
	Fe	iron	
	Co	cobalt	
	Cu	copper	
	Zn	zinc	

The three elements of the left-hand list, nitrogen, phosphorus and sulphur, are absolutely essential for the functioning of every cell. Nitrogen and sulphur, like carbon, hydrogen and oxygen, occur in the environment in gaseous form. The elements in the middle list are ubiquitous in living organisms while those in the right-hand list are essential either for the majority of organisms or for important groups of organisms. In each list the elements are ordered in terms of increasing atomic number, and not their relative abundance. Sometimes, nitrogen is grouped along with carbon, hydrogen and oxygen because it is an integral part of the structure of some vitally important biological substances. In addition, nitrogen makes a significantly greater contribution to the dry weight of microorganisms, notably the bacteria, than to plants and animals.

Even though the elements of the two right-hand lists are required in comparatively small amounts by plants and animals, considerable differences occur between them in the concentrations in which they are normally present. Whereas nitrogen normally contributes over one per cent to the dry weight of a plant, copper and iron are present in very much lower concentrations, probably accounting for less than 0.001 per cent of the weight. On the basis of the concentrations in which they are normally required, elements are referred to either as *macro*nutrients (and we can include carbon, hydrogen and oxygen in this category), or *micro*nutrients (or *trace elements*). However we should not equate amount with importance as far as an organism's metabolism is concerned.

Because a few elements are required by some types of organism but not by others, and because there is some dispute as to whether certain elements are really essential, you may well find some slight variation between lists of essential elements in different references. A detailed chemical analysis of biological tissue may well reveal the presence of a number of elements which are not essential for metabolism, and by definition cannot be considered nutrients. However, they can be taken up by organisms from their environment or consumed in the diet. Some of these elements, such as lead and cadmium, are very toxic even at low concentrations, which is why their discharge into the environment is a matter of considerable concern. However, we should not assume that all nutrients are benign in whatever concentrations they occur. Copper and zinc exemplify elements required in tiny amounts by almost all organisms but which become very toxic at high concentrations. The devastated vegetation typically found around copper and zinc smelters and the practically lifeless streams that emanate from some mine wastes demonstrate this point.

Elements in the Earth's crust

The following list shows the approximate percentage contribution by weight of the eight most abundant chemical elements in the rocks and minerals at or near the Earth's surface.

		%
O	oxygen	47
Si	silicon	28
Al	aluminium	8
Fe	iron	5
Ca	calcium	4
Na	sodium	3
K	potassium	3
Mg	magnesium	2

Clearly, the ratios in which elements occur within living tissue do not reflect their relative abundance in the Earth's crust. Silicon for example, the second most abundant element in crustal rocks is not used by all types of organism. Aluminium, the third most abundant element, is not required by any organism, and in solution is quite toxic. Carbon and nitrogen, key biological elements, do not appear in this list: they are in fact ranked seventeenth and twenty-fifth respectively (although carbon is locally abundant, as carbonates, and also in coal measures).

Molecules and compounds

Atoms combine with each other in characteristic ratios to form *molecules*. Formally defined, a molecule of a particular substance is the smallest part of that substance which can exist independently and still retain the properties of that substance. Some molecules are made up of atoms of a single element. Molecular oxygen and nitrogen, for example, both comprise a pair of atoms, in which form they are symbolized as O_2 and N_2 respectively. Other molecules are made up of two or more elements.

The term *chemical compound* is used for a substance that is composed of atoms of more than one element in fixed proportion. Numerical subscripts placed after the element symbol denote the ratios of the different types of atoms. Thus a molecule of carbon dioxide (CO_2) is composed of twice as many oxygen atoms as carbon atoms, a molecule of ammonia (NH_3) is made up of nitrogen and hydrogen atoms in a 1:3 ratio and a molecule of glucose ($C_6H_{12}O_6$) consists of six atoms of carbon, twelve atoms of hydrogen and six atoms of oxygen. These short-hand representations are examples of *molecular formulae*. The numbers of atoms in a molecule of glucose can be divided by six to derive the *empirical formula* (CH_2O) which simply shows the proportions in which the different atoms occur.

A molecular formula does not show the spatial arrangement of atoms in a molecule; for this the structural formula is required. We shall meet a number of examples of structural formulae a little later. In the case of some biologically important substances, more than one substance may share the same molecular formula, so depicting a compound in structural terms is very useful.

Chemical bonds and ions

The forces which attract and bind atoms to each other are referred to as *chemical bonds*. Of particular importance in understanding bonding processes is the behaviour of electrons, the negatively-charged particles in motion within shells around atomic nuclei. Recall that electrons do not necessarily move in circular orbit, although it is often convenient to represent them in this way: rather, an electron moves within a particular volume, known as an orbital, with a characteristic geometry.

The atoms of each element have a characteristic number of electron shells, ranging from one to six. The innermost shell (and the single shell in the case of hydrogen and helium) is said to be *complete* when it has two electrons. Other shells are complete with a larger number of electrons, usually eight. In general, atoms behave in such a way as to attain complete outer shells. Now, if we were to check the number of electrons in the shells of all the naturally occurring elements, we would find that very few have complete outer shells. However, atoms can attain this state, either by sharing electrons with other atoms, by releasing electrons, or by acquiring additional electrons from other atoms. These processes result in important types of chemical bonds that bind atoms together to produce molecules.

The first type of bond we consider, called a *covalent* bond, forms stable molecular structures; it is the most important way in which atoms are bound to each other in the substances that interest us. Covalent bonding involves the sharing of electrons in the outer shells of atoms. Oxygen, which exists primarily as molecules made up of two oxygen atoms (O_2) provides a simple example of covalent bonding. An oxygen atom has six electrons in its outer shell, two less than the 'complete' number: however, by sharing two electron pairs, each oxygen atom of a pair can effectively have eight electrons in its outer shell (Fig. 2.2). The gas methane (CH_4) is also held together by covalent bonds, but in this case single electron pairs are shared between each of the four hydrogen atoms and the central carbon atom (Fig. 2.2).

The sharing of electron pairs between atoms is represented by dashes, as shown for oxygen and methane in Fig 2.2; the number of shared electron pairs, i.e. the number of bonds, being shown by the number of dashes. We shall represent chemical structures in this way when we introduce the major classes of substances in living organisms. Carbon atoms combine readily by single, double or triple bonds, with each other and with oxygen, hydrogen and nitrogen. Thus carbon can form a huge number of different compounds.

Also very important is the *ionic* bond. Recall that an atom with an equal number of positively-charged protons and negatively-charged electrons is electrically neutral. But if a neutral atom loses one or more electrons it becomes positively charged; conversely, if a neutral atom gains an electron it becomes negatively charged. The process of losing or gaining electrons is called *ionization*, and the entities which result are *ions*. Positively-charged ions are called *cations*, negatively-charged ions

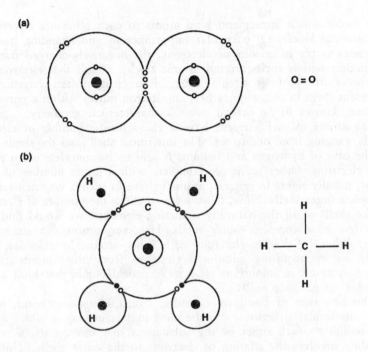

Fig. 2.2 The principle of the covalent bond. (a) Molecular oxygen (O_2): two pairs of electrons are shared between two oxygen atoms so that each atom effectively has eight electrons in its outer shell. This is a double bond and is represented by two dashes. (b) Methane (CH_4): a carbon atom is bound to each of four hydrogen atoms by the sharing of a single electron pair, thus the single dashes in the shorthand structure shown alongside.

are called *anions*. Ions are represented by the appropriate element symbol (or symbols, because more than one element may be involved), plus the appropriate sign and an indication of the number of excess charges. Here are some examples, all of which are very important to the biosphere.

	Cations		Anions
K^+	potassium	Cl^-	chlorine
NH_4^+	ammonium	NO_3^-	nitrate
Mg^{2+}	magnesium	SO_4^{2-}	sulphate

The simultaneous donation and acceptance of electrons between atoms gives rise to ions carrying opposite electrical charges. The mutual attraction between oppositely charged ions is the basis of the ionic bond. (Also in fact, some ions can become further charged by losing or gaining more electrons.)

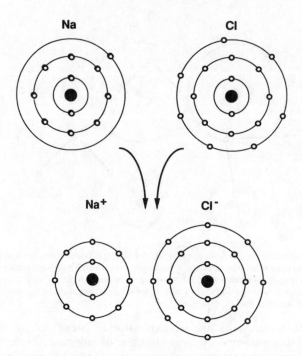

Fig. 2.3   The principle of the ionic bond. In the two top diagrams 'lone' atoms of sodium
(Na) and chlorine (Cl) are shown. By losing the sole electron in its outer shell,
sodium has a new outer shell with eight electrons and becomes positively
charged; chlorine, by gaining this electron, acquires an outer shell of eight
electrons and becomes negatively charged. The attraction between oppositely
charged ions is the basis of an ionic bond.

A familiar substance in which atoms of different elements are bound
by ionic forces is common salt, sodium chloride (Fig. 2.3). A sodium
atom has one electron in its outermost shell and eight electrons in the
shell beneath. A chlorine atom has seven electrons in its outermost shell.
Therefore by donating its sole outer electron to a chlorine atom, a
sodium atom becomes a cation with a new outermost shell containing
eight electrons. The chlorine atom simultaneously acquires a complete
outer shell of eight electrons and becomes negatively charged. The
sodium and chlorine ions therefore become bound together to form a
fairly stable entity.

Ionic bonds can involve more than two atoms. Calcium chloride
($CaCl_2$) is made up of calcium and chlorine atoms in a 1:2 ratio.
There are two electrons in the outer shell of a calcium atom and, like
sodium, eight in the shell beneath. Calcium therefore achieves a stable
state by donating the two outer electrons. The number of protons in the
nucleus then exceeds the number of electrons by two; so calcium ions
carry a double positive charge, represented as Ca^{2+} or Ca^{++}. When

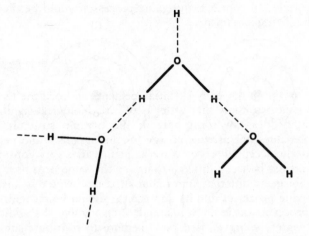

Fig. 2.4   The principle of the hydrogen bond. A hydrogen bond arises from slight
differences in charge between a hydrogen and a neighbouring (usually) oxygen
or nitrogen. Within water, each oxygen is bound covalently to a pair of
hydrogens (heavier lines), but also to another hydrogen by a weaker, hydrogen
bond (dashed lines).

dropped into water, salts like sodium chloride 'break up', or *dissociate*,
into ionized components. A large number of substances behave in this
way and many elements move through the biosphere principally in an
ionized state.

One other type of chemical bond we will encounter is the *hydrogen*
bond. Hydrogen bonds involve the sharing of a hydrogen atom between
two other atoms, usually of oxygen or nitrogen. The hydrogen atom is
covalently bonded to one of these atoms, but because it shares its electron, the hydrogen assumes a slight positive charge. Therefore the
hydrogen attracts atoms which are electrically negative in other
molecules. Hydrogen bonds, which are much weaker than either covalent
or ionic bonds, are largely responsible for the unusual properties of
water, which of course is essential for all life processes. The position of
the hydrogen bonds in water are shown in Fig 2.4.

Organic and inorganic

The terms 'organic' and 'inorganic' are in such common use that further
comment may seem unnecessary. In fact though these terms are difficult
to define precisely. In general terms, 'organic chemistry' is concerned
with covalently-bonded, carbon-based compounds, and not just natural
carbon compounds. 'Inorganic chemistry' deals with substances in which
carbon is absent or not prominent. 'Organic' is used also, however, in
the sense of 'biological'; processes may be described as organic when they
are brought about by living organisms; organic matter refers to material

of biological origin. In contrast, inorganic processes would be those that do not involve living organisms.

Water

While discussing the biologically essential elements we used the term 'dry weight'. This is not to imply that water is of little biological significance. However, water does not form part of the organic structure of an organism; water does not provide energy for the organism and neither is it a nutrient in the accepted sense. Water content varies very considerably between organisms, between different parts of the same organism and in the same organism at different times. But of course water is absolutely essential for living processes and its importance should be recognized. All biochemical processes occur in water and often involve the addition or formation of water; water is used as a medium to transport substances around the body and to remove waste and unwanted substances; in addition, the evaporative loss of water from an organism brings about surface cooling, an important mechanism for dissipating heat in hot conditions.

Major classes of biochemical substances

In introducing the major biochemical substances we are faced with something of a paradox. On the one hand, organisms collectively synthesize a vast number of different chemical compounds, a point that should be emphasized. On the other hand, we also want to emphasize the point that the chemical composition of organisms is a unifying characteristic of the biosphere. The fact is that despite the biochemical diversity that undoubtedly exists, most of the weight of any organism is made up of compounds belonging to just a few major classes of chemical substance, and moreover, certain types of chemical molecule are present in all living cells. Learning some essential features of the major biochemical substances is therefore relatively straightforward, particularly as many of the terms are in everyday use.

Proteins

Proteins are so central to living processes that we can regard them as synonymous with life itself. Proteins are present in every living cell where they perform a wide range of vital roles. Most importantly we associate proteins with their role as *enzymes*. Essentially, enzymes are substances which act as catalysts, or 'accelerators', for chemical reactions. Virtually every aspect of a cell's metabolism is controlled by enzymes. Differences between organisms are closely associated with differences in the enzymes they produce.

A very common type of enzyme-mediated reaction in living cells involves the addition of water to the products of a substance as it is broken down, thus:

$$A + H_2O \rightarrow B(OH) + C(H)$$

In this type of reaction, termed *hydrolysis*, the molecule represented by A is split, under the control of a specific enzyme, into two smaller molecules, B(OH) and C(H). During this process hydrogen and oxygen atoms from the water are added somewhere to the products of the reaction. The reverse type of reaction, involving the elimination of water during the linking of two molecules, is also very common. It is referred to as a *condensation* reaction. In the case above, A would be produced from B(OH) and C(H), with the elimination of a water molecule. Importantly, enzymes are not changed or 'used up' during the reactions they control.

Enzymes are fundamental to life, and we will refer to them frequently in this book. The digestion of food provides a very familiar example of enzyme activity. Within cells, most biochemical transformations involve a sequence of chemical reactions, each one controlled by a specific enzyme, and each one resulting in a fairly small change in the structure of the chemical molecules involved. A series of chemical reactions of this sort is known as a *biochemical* or *metabolic pathway*. Essentially, it is the biochemical pathways that determine how an organism 'works'.

In addition to their role as enzymes, proteins perform a variety of other vital functions. Antibodies, which provide our bodies with protection against foreign substances, are mainly protein; muscle fibres are rich in protein (which is why animal flesh is such an excellent source of dietary protein); hoof, hair, nail and cartilage are also largely protein. Protein molecules are not normally used as a source of energy for living organisms, but when the principal types of energy storage compound are in short supply protein may be metabolized for this purpose.

Chemically, proteins are very large, very complex, molecules. However, they are made up of several much smaller and simpler substances called amino acids which are linked together. (The general term for a long chain of chemical units of similar type is *polymer* and the process of linking the component units together is *polymerization*.)

The general formula for an amino acid is shown in Fig. 2.5. (Notice here that only some of the chemical bonds between atoms are shown: this is standard practice when representing molecular structures diagrammatically.) The essential features of an amino acid are, first, an amino (NH_2) group and, second, a carboxyl, or 'acidic', (COOH) group. Amino acids are linked together by the amino group on one molecule and the carboxyl group on another: this is an example of a condensation reaction (it results in the formation of a water molecule). Strictly, proteins are made up of amino acid *residues*, because in the linking of two amino acids, to produce a *peptide* bond (Fig. 2.5), two atoms of hydrogen and one of oxygen are eliminated. The term polypeptide refers to any string of linked amino acid residues. The R in our diagram

(a)

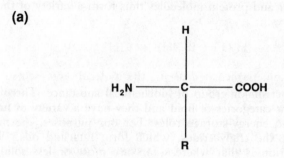

(b)

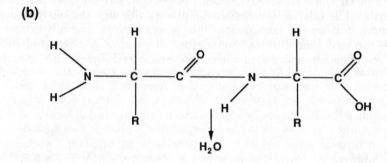

Fig. 2.5 (a) Basic structure of an amino acid. R represents a group whose structure varies between the twenty or so different amino acids. (b) Arrangement of atoms in two amino acid residues linked by a peptide bond between the carboxyl (COOH) and amino (NH$_2$) groups; water is eliminated so this is an example of a condensation reaction. A protein consists of hundreds or thousands of linked amino acid residues and is usually folded and twisted into a particular shape.

represents a chemical group which varies between amino acids and which distinguishes the amino acid. In the simplest amino acid, glycine, it is a single hydrogen atom, but in the other amino acids, R is a larger chemical group. Notice the presence of nitrogen, which contributes around fourteen per cent by weight to protein. Sulphur and phosphorus are often present in protein as they are contained in certain amino acids.

Just twenty amino acids are commonly found in proteins, but this number is quite sufficient to allow an enormous variety of proteins. Proteins differ most obviously in the sequence, and number, of amino acids: forty is around the minimum number but there may be more than a thousand. Even if all proteins had a fixed length of one hundred amino acids, with twenty different amino acids the potential number of proteins would be a staggering 20^{100}! The sequence of amino acids is known as the *primary* level of organization. However, protein structure is rather more complex: cross-linkages between side groups of amino acids join polypeptides together and protein molecules thus form a variety of three-dimensional structures.

Lipids

The lipids are another major group of biochemical substance. There are a number of different categories of lipid and they have a variety of functional, structural and energy-storage roles. For our purposes, the most important lipids are the triglycerides, which form fats and oils. (The conventional distinction is that whereas fats are more or less solid at 'normal' temperatures (around 20°C), oils are liquid, although frequently the term fat is used to refer to both.) In general, fats are associated with animals while oils are synthesized by plants, although the oils produced by many fish are an exception to this generalization. The oils extracted from the seeds and fruits of some types of plant, e.g. olives and corn, are very familiar because of their commercial and dietary importance.

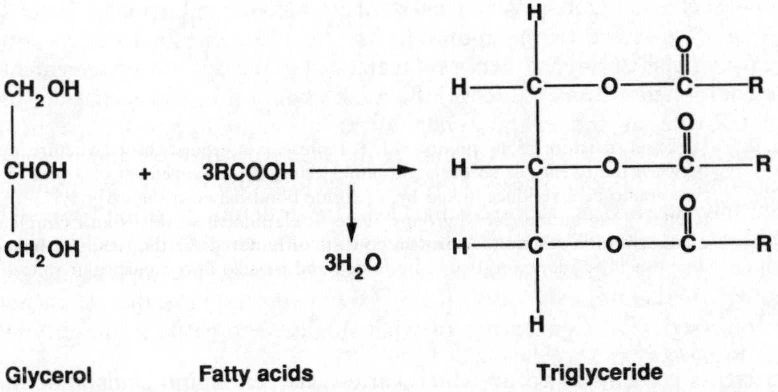

Glycerol   Fatty acids   Triglyceride

Fig. 2.6   The formation of a triglyceride. Glycerol and fatty acids are linked by condensation reactions to produce triglycerides. The spatial arrangement of atoms is shown for part of the resulting triglyceride molecule. R represents a chain of carbon atoms whose length and number of associated hydrogen atoms determine the type of fatty acid, and hence the type of fat.

A triglyceride is made from one molecule of glycerol and three molecules of fatty acid, which may or may not be of the same type (Fig. 2.6). Again, water molecules are eliminated when bonding occurs, leaving residues of both glycerol and the fatty acids. The R group on the fatty acids in Fig. 2.6 represents a chain of carbon atoms, most commonly between four and twenty-four, to which are attached hydrogen atoms. Variation in the type of fatty acid gives rise to different triglycerides, and the fat or oil from a particular tissue may contain a mixture of triglycerides.

Fats are not the only type of lipid of importance. The waxes, which comprise fatty acids and alcohols, form surface films on plant leaves and fruit and on animal fur and feathers. Waxes confer some protection, particularly against desiccation and wetting. The phospholipids, so called on account of their phosphorus component, are vital components of the membranes that surround plant and animal cells. The steroids which include a number of animal hormones, notably the sex hormones, are also classed as lipids.

Carbohydrates

Carbohydrates are universally present in living organisms. Several types of carbohydrate occur in nature, but they all share certain key features. The building blocks of carbohydrates (analogous to the amino acids of proteins) are monosaccharide units. The simplest carbohydrates are called monosaccharides and they have between two and seven carbon atoms, with six the most common number. We used the six-carbon monosaccharide glucose when introducing molecular formulae (page 11). Actually, several different monosaccharides share the molecular formula $C_6H_{12}O_6$, the difference between them being the spatial arrangement of the component atoms. Glucose, the most common monosaccharide, plays a vital role in the energy metabolism of many organisms. Fructose, another monosaccharide, is found in the nectar of flowers and is a major ingredient of honey. These, however, are just two of a large number of naturally occurring monosaccharides. In aqueous solution (i.e. when dissolved in water) monosaccharides exist in ring form, as shown for glucose in Fig. 2.7. Actually, only one form, called alpha-glucose, is shown; on beta-glucose the O and OH (hydroxyl) groups on carbon-1 are reversed. The significance of what might seem to be a fine difference will be apparent shortly.

Disaccharides consist of two monosaccharide residues, linked by a condensation reaction between OH groups (Fig. 2.7). The 'sugar of commerce' is the disaccharide sucrose, one molecule of which consists of a glucose unit and a fructose unit linked together. For commercial purposes, sucrose is extracted principally from two plant types, sugar beet, a bulbous-rooted plant of temperate regions, and sugar cane, a grass of tropical origin. Lactose, which is present in human milk, is a disaccharide made from glucose and another monosaccharide, galactose. Carbohydrates with three, four and five monosaccharide units are also found naturally, and like monosaccharides and disaccharides, are known as sugars.

Other biologically important carbohydrates are made up of a large number of linked monosaccharide units. The most abundant of these *poly*saccharides are based exclusively on glucose units. (Note the contrast with proteins in which the principal source of variation is the sequence of

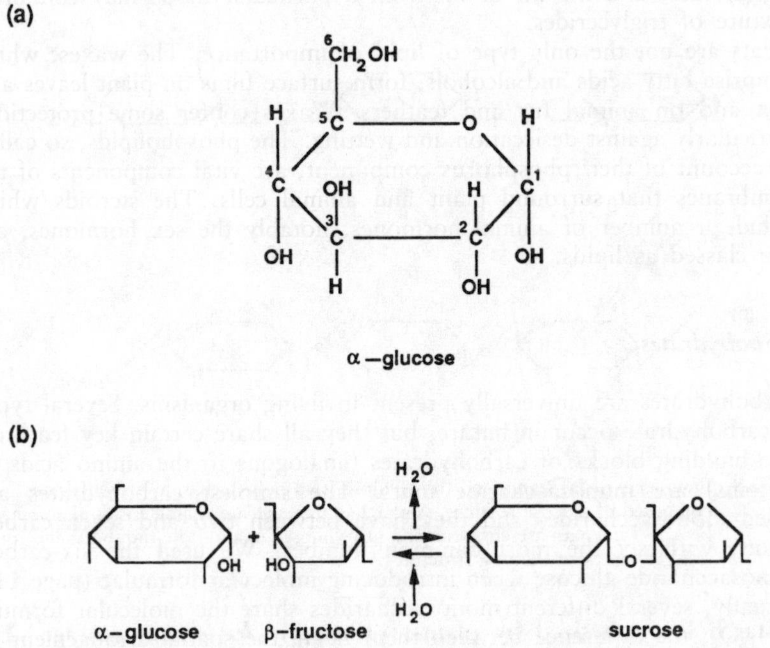

(a)

α —glucose

(b)

α—glucose β—fructose   sucrose

Fig. 2.7 (a) The ring structure of glucose, a very common monosaccharide carbohydrate. The numbering system shown for the carbon atoms is in general use. The alpha form is depicted; in the beta form the positions of the oxygen and hydroxyl (OH) group at carbon-1 are reversed. (b) The disaccharide sucrose, composed of one alpha-glucose residue and a beta-fructose residue linked at carbons 1 and 4. Compare the shorthand representation of the glucose unit with that in (a). Note the differences in ring structure between glucose and fructose.

different amino acid building blocks.) The two most abundant glucose-based polysaccharides are starch and cellulose, but they differ significantly in their chemical properties and in their biological roles. Starch (Fig. 2.8) is a mixture of two glucose-based polymers. In one polymer, glucose residues are linked linearly by alpha 1–4 linkages (i.e. between hydroxyl groups at carbon-1 on one molecule and carbon-4 on the other, alpha-glucose, molecule). In the other polymer of starch, most glucose units are linked linearly in the same way, but in addition a small proportion of the glucose residues are linked to each other at carbons-1 and 6, which gives the substance a branched structure.

In contrast to starch, cellulose is based on beta-glucose residues (Fig. 2.8). The principal points of difference between these extremely important carbohydrates are the form of glucose and the alignment of bonds between glucose units. As a consequence starch and cellulose are very different substances and, very importantly, different enzymes are needed to hydrolyse them, i.e. break them down to their component sugar units.

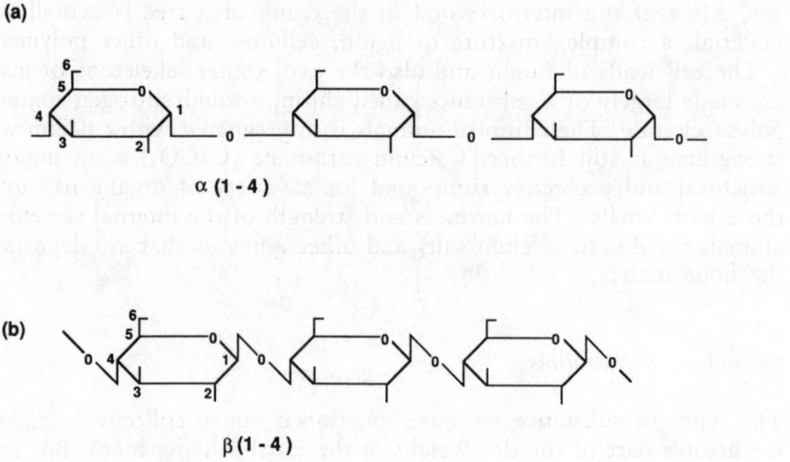

Fig. 2.8   Fragments of (a) starch and (b) cellulose, two polysaccharide carbohydrates composed of glucose residues. Alpha 1–4 linkages are characteristic of starch, beta 1–4 linkages are characteristic of cellulose. Again, note the shorthand form used to represent the ring structures.

Starch is synthesized by plants, principally for energy storage. The concentration of starch varies between different parts of the same plant, and also between plant types. The starch content of seeds is usually quite high. Starch is also stored in the underground organs of many perennial plants, particularly during times of the year when conditions are unfavourable for growth. Starch, derived from cereal grains and certain root and tuber plants such as potatoes and yams, makes up a large proportion of the energy intake for an overwhelming majority of the human population.

Cellulose is also manufactured by plants, but unlike starch, it is a structural carbohydrate. It is found principally in plant cell walls which enable the cell to resist swelling from internal water pressure. Cellulose also considerably strengthens the plant structurally, enabling some plants to reach considerable heights. As a plant ages, more and more cellulose is laid down in its cell walls. Cellulose is the most abundant substance in the Earth's biota.

Glycogen, another polysaccharide based on glucose units, deserves mention because it is the major energy-storage carbohydrate of mammals. In humans, glycogen is found chiefly in muscle cells and in the liver.

Lignin and other structural substances

Lignin is a chemically complex substance, synthesized by plants and laid down in cells walls, along with cellulose, in increasing amounts as plants age. Most of the interior wood in the trunk of a tree is actually dead material, a complex mixture of lignin, cellulose and other polymers.

The cell walls of fungi, and also the exo- (outer) skeletons of insects, are made largely of a substance called chitin, a tough nitrogen-containing polysaccharide. The chitin of animals is impregnated with calcium which strengthens it still further. Calcium carbonate ($CaCO_3$) is an important structural and protective compound for a variety of organisms, notably those with 'shells'. The hardness and strength of the internal skeletons of animals are due to calcium salts and other minerals that are deposited in the bone matrix.

Secondary metabolites

The types of substance we have mentioned above collectively make up the greater part of the dry weight of the Earth's living biota. But collectively, organisms also synthesize a wide range of substances, referred to as *secondary metabolites*. These are chemical compounds that are not involved directly in the organism's metabolism, but do influence its relationships with other organisms. In this category are substances produced by plants that serve as feeding deterrents and the toxins produced by fungi that inhibit growth of bacteria.

By definition, secondary metabolites are 'bioactive', a property that may be put to good use, perhaps for medicinal purposes or for pest and disease control in crops and livestock. In fact many of the medicinal drugs in common use were extracted originally from plants or other organisms. Extinction of species then brings with it the loss of potentially valuable biochemical resources, so providing a quite pragmatic argument for the conservation of organisms and their habitats.

The nucleic acids

The nucleic acids practically define life as we understand the term. We introduce them here, rather than at the beginning of this section, simply because the topics that follow build logically on some appreciation of what these substances are and how they function. There are two nucleic acids to consider, deoxyribonucleic acid (DNA) and ribonucleic acid (RNA), both of which are present in every living cell. Essentially, the DNA carries all the information, in chemically coded form, necessary for the functioning and the development of the cell while RNA serves to translate this information into action, principally through the synthesis of proteins. As DNA is passed on from generation to generation, it provides the basis for the inheritance of characteristics and the continuity of life.

Chemically, DNA is made up of units called nucleotides (Fig. 2.9). Each nucleotide is itself formed from smaller units, a five-carbon sugar, a phosphate group and one of four types of nitrogenous base. The four different types of nitrogenous base are adenine, thymine, guanine and

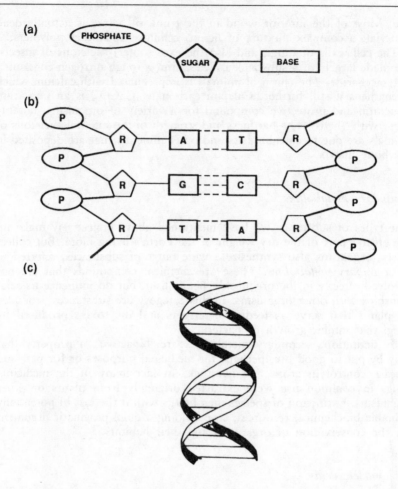

Fig. 2.9 Structure of DNA. (a) The building blocks of a nucleotide; (b) A fragment of
DNA showing its double-stranded structure: (P) phosphate group; (R) ribose, a
5-carbon sugar; the nitrogenous bases are (A) adenine, (T) thymine, (C) cytosine
and (G) guanine. Dashed lines denote the hydrogen bonds linking base pairs,
two for adenine – thymine, three for cytosine – guanine. Heavier lines denote
the bonds linking adjacent nucleotides. (c) The double helix structure of DNA;
each 'rung' comprises a pair of bases linked by hydrogen bonds, the 'sides' are
made up of linked sugars and phosphates.

cytosine. The nucleotides of DNA are linked by chemical bonds between
the sugar unit of one nucleotide and the phosphate group of the adjacent
nucleotide. DNA exists as a double-stranded molecule, with the two
strands running in opposite directions (Fig. 2.9). The nitrogenous bases
of the two strands face each other and are linked to each other by
hydrogen bonds. Very importantly, adenine only links with thymine
while guanine only links with cytosine. The base pairs that link with

each other in this way are termed *complementary*. The DNA molecule is twisted around its own axis to form a structure known as a *double helix* (Fig. 2.9). An analogy sometimes used to assist appreciation of DNA structure is that of a ladder with rope sides and rigid rungs. Chemically, each side of the 'ladder' is made of the sugar and phosphate groups while each 'rung' consists of a complementary pair of nucleotide bases. A spiral staircase provides another analogy. The elucidation of the structure of DNA, in the early 1950s, is generally acknowledged to be one of the most important advances ever in the history of the life sciences, and it led to the award of Nobel prizes to three scientists, James Watson, Francis Crick and Maurice Wilkins.

The vital point about DNA is that the sequence of nucleotide bases forms a chemical code, known as the *genetic code*, that is universal for all forms of life on this planet. A particular sequence of three adjacent nucleotide bases, known as a *triplet*, specifies a particular amino acid. The order in which amino acids are linked together (by peptide bonds), and hence the type of protein produced, is directly related to a sequence of base triplets. Actually, there are sixty-four different base triplets (they can be thought of as 'words'), many more than the number of amino acids, but the same amino acid may be coded for by more than one triplet. And some triplets code for the 'stopping' of amino acid sequencing. So, as an example (using just the first letter of the nucleotide base), the sequence of triplets CCA-AAG-CGG specifies a different amino acid sequence to the sequence of triplets AAC-GCC-ACG. As the difference between types of protein is primarily a function of difference in amino acid sequence, it is essentially the variation in nucleotide sequences that is the basis for variation between organisms.

The DNA does not do the 'work' of protein synthesis however, that is the function of the other nucleic acid, RNA. In eukaryotic cells (see page 4), the DNA is located within the nucleus, but protein synthesis takes place outside the nucleus, in the cellular cytoplasm.

Now RNA and DNA differ in certain respects. First, in RNA the nitrogenous base uracil takes the place of thymine: uracil is therefore the complementary base for adenine. Second, the sugar units of DNA have one less oxygen atom (thus *de*oxyribonucleic acid) than those of RNA. In addition, RNA is a single-stranded, not a double-stranded, molecule. It is the role of RNA molecules to 'read' the information stored in code on the DNA, to carry this information to the sites in the cell where proteins are synthesized, and to orchestrate the sequencing of amino acids in the correct order for the production of proteins as specified by the piece of DNA that is 'read'. Three different types of RNA, with different names, are involved in these processes.

Protein synthesis

Protein synthesis (Fig. 2.10) is generally described as a two-stage process; the first stage is called *transcription*, the second stage is called *translation*.

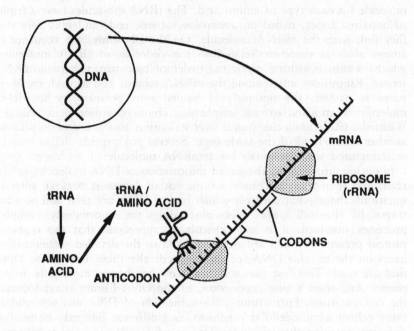

Fig. 2.10 Schematic representation of polypeptide formation at the ribosomes in a eukaryotic cell. Information stored on DNA in the nucleus is carried into the cytoplasm by complementary strands of messenger RNA. Within the cytoplasm, specific transfer RNA molecules pick up the amino acids and hand them on to the rRNA of the ribosomes, where they are joined to a lengthening peptide chain. Note the characteristic shape of a tRNA molecule. The sequence of amino acids is determined by the order of base triplets (codons) on the messenger RNA strand.

Transcription involves the 'reading' of the chemical code on the DNA. This requires the 'unzipping' of the DNA molecule and the synthesis of a strand of *messenger* RNA (mRNA) of complementary base sequence. A special enzyme is involved in this process: it moves along the DNA strand, identifies the nucleotide base it encounters, and adds the appropriate complementary base to produce the mRNA strand. The mRNA strand then migrates to the cytoplasm.

The translation stage involves the ordering of amino acids as specified by the sequence of triplets on the strand of mRNA. A sequence of three bases on a strand of mRNA is called a *codon*. The site of protein synthesis is the *ribosome*, which is made up of another type of RNA, *ribosomal* RNA (rRNA), as well as protein. Ribosomes, which occur in a group known as a *polysome*, connect with the mRNA (Fig. 2.10). Molecules of another type of RNA, *transfer* RNA (tRNA), which have a characteristic shape (Fig. 2.10), pick up amino acids in the cytoplasm and convey them to the ribosomes. There is a specific type of tRNA molecule for each type of amino acid. The tRNA molecules have a triplet of unpaired bases, called an *anticodon*, at one end, and it is here that they link with the mRNA molecule. On the ribosomes the sequence of amino acids is therefore related to the ordering of tRNA molecules, which in turn is controlled by the order of base triplets on the mRNA strand. Ribosomes move along the mRNA strand, and as each triplet of bases is passed, the amino acid hauled into position by its tRNA molecule is attached to the lengthening chain of amino acid residues. When the tRNA molecule hands over its amino acid it is free to pick up another amino acid of the same type. Several poylpeptide chains may be manufactured simultaneously on a mRNA molecule.

In outline, this is how the coded information of DNA molecules is first read, and then translated into action, but of course it begs all sorts of questions about what controls which bits of DNA are read and at what stages of the cell's life. Notwithstanding the enormously complex processes involved, it is not difficult to appreciate that the types of protein present in a cell are closely related to the detailed sequencing of bases on the cellular DNA, or more specifically those bits of the DNA that are read. This last caveat is necessary because in most cells only a proportion, often a tiny proportion, of the DNA library is read during the cell's lifetime. Furthermore, the same bits of DNA are not read in every cell of a multicellular organism, and different bits may be read at different times in the cell's life. This is why cells, and the tissues and organs they comprise, perform different functions.

The amount of information stored in the DNA of a cell is enormous. It is estimated that within the nucleus of a human cell there are a staggering three thousand million (i.e. three billion) bases, and within a comparatively simple bacterium the number is likely to be a few million. However, as just pointed out, only a fraction of a cell's DNA may be actually used. In summary then, developmental and functional processes in cells and organisms are determined ultimately by the information stored on molecules of DNA. As we discuss in the following chapter, DNA is transmitted from parent to offspring during reproduction thus ensuring the continuity of life.

3 Genes, chromosomes and cell division

A *gene* can be regarded as a sequence of nucleotides from a DNA molecule that codes for a particular protein, although as discussed towards the close of the previous chapter, the job of protein synthesis is carried out by RNA. The same genes are present in every cell of a multicellular organism and they collectively constitute that organism's *genome*. The number of genes per cell varies enormously between different types of organism; for humans it is estimated to be over one hundred thousand. There are differences to some extent between cells of the same organism as to which bits of the genome are 'read'. Heart and brain cells from the same individual share the same genetic information; some essential cellular processes are common to both, but each in addition carries out specialized functions and they appear different under the microscope.

Chromosomes

The genes of eukaryotic cells are carried on thread-like structures called *chromosomes* which are located within the cell nucleus. A eukaryotic chromosome comprises a single DNA molecule together with some protein. Chromosomes are commonly likened to rods, but it is only when a cell nucleus is dividing that they appear as such. In prokaryotic cells, which have no nucleus, the DNA molecule appears as a loop, attached at one point to the cell membrane, and there is no associated protein.

In eukaryotic cells, the chromosomes occur as pairs, known as *homologous pairs*. With the exception of the sex-determining chromosomes which are characteristic of most animals and some plants, each chromosome, i.e. each homologue, of a pair is effectively identical in appearance. However, as we shall see, at some stage in the life history of eukaryotic organisms, only one, not both homologues, are present. For simplicity we can regard each chromosome of a pair as a 'version' of that chromosome. Essentially, each chromosome of a homologous pair carries genes that code for the same general structures and functions, and the genes are ordered in the same way, i.e. at the equivalent location, on each chromosome of the pair. There are usually alternative 'versions' of each gene, referred to as *alleles*, and different alleles may well occur on the two chromosomes of a homologous pair. In practice most traits are determined by combinations of genes.

The two sex chromosomes of an organism, unlike the other chromosome pairs, may differ considerably in appearance. Either male or female individuals (it depends on the kind of organism) carry a pair of identical sex chromosomes (in humans it is the female), whereas individuals of the other sex carry one of each type of sex chromosome. Nevertheless, even when both types are present in the same cell they can be regarded as a pair. So now when we use the term 'chromosome pairs', the sex chromosomes are included. Each kind of organism has a characteristic number of chromosome pairs per nucleus. Humans have twenty-three chromosome pairs and forty-six chromosomes in all (but there are twenty-four 'types' of human chromosome simply because of the two sex chromosomes).

If an organism has different alleles at the equivalent location on homologous chromosomes, it is said to be *heterozygous* for whatever trait is determined by those genes. But if identical alleles are present on a chromosome pair the individual is described as *homozygous* for the specified trait. Expressed alleles are described as *dominants*, those that are not expressed when present are termed *recessives*.

Diploid and haploid cells

A cell, or nucleus, with the full complement of chromosome pairs is described as *diploid*; a cell or nucleus with only one chromosome of each pair is described as *haploid*. The diploid number is symbolized by $2n$, the haploid number by n. So, for human cells, $2n = 46$ and $n = 23$. The cells of most familiar plants and animals are diploid, i.e. both homologues of each chromosome are present. But there are exceptions. The 'bodies' of mosses, those small seedless plants frequently found in damp situations, are composed of haploid cells, although they do in fact produce smaller entities with diploid nuclei. We shall return to this topic a little later, but first we need to consider how cells divide, a process which is central to growth and reproduction.

The division of cells

Growth and reproduction clearly require cell multiplication. Although it may seem a little illogical, the process by which this is brought about is known as *cell division*. The reason for this is that the mechanisms involved in cell reproduction are more faithfully represented by this term.

Reproduction of a prokaryotic cell, i.e. a bacterium, occurs by *binary fission* (Fig. 3.1). The cell does indeed appear to divide. A copy is made of the single DNA molecule quite early in the cell cycle; later the DNA copies separate, and a membrane is completed around each so that two, genetically identical, independent cells are formed.

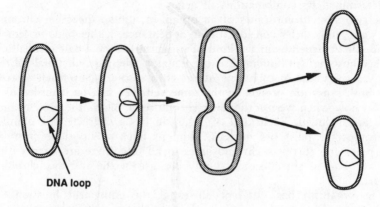

DNA loop

Fig. 3.1 The reproduction of a prokaryotic cell. The DNA, which exists as a single molecule, is copied and later the cell divides.

For eukaryotic organisms, cell division is a little more complex. In describing this process it is necessary to concentrate on the chromosomes, because they carry the genetic information that determine what the new cells will be like. Strictly, we are concerned with the division of nuclei because nuclear division does not necessarily involve the formation of two new cells. (In fact there are certain times during the life cycle of some kinds of organism when the formation of multinucleate cells is normal.) Henceforth, when we use the term nuclear division, it is with the understanding that cell division normally accompanies nuclear division, but that nuclear division can occur independently of cell division. There are two types of nuclear (or cell) division, known respectively as *mitosis* and *meiosis*. We shall deal with each in turn, emphasizing the behaviour of the chromosomes.

Mitosis

A mitotic division gives rise to two new nuclei, each with the same number of chromosomes as the original nucleus and of identical genotype. Mitosis must therefore involve the copying of the DNA in the original nucleus. Before it is copied, the DNA molecule is 'unzipped', and then two new strands of DNA with complementary base sequences are synthesized. Thus two identical DNA molecules are produced. The behaviour of chromosomes during a mitotic division is shown in Fig. 3.2. The nuclear membrane surrounding the chromosomes disappears and chromosomes condense so that it is possible to recognize the homologous pairs. The genetic material on each chromosome has been copied, and each chromosome now comprises two distinct entities, known as

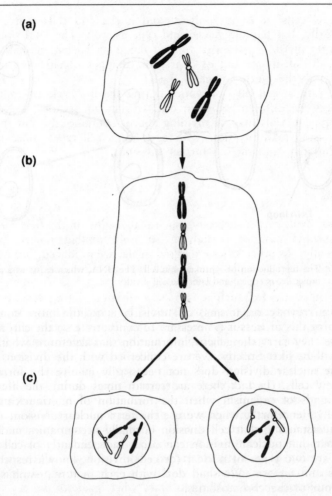

Fig. 3.2 The behaviour of chromosomes during mitosis. Here the diploid (2*n*) number is four. Chromosomes have duplicated (a), but remain joined at the centromeres. They then line up (b) prior to separation. Chromatids separate in such a way that each new nucleus contains an identical set of genetic material (c).

chromatids. The two chromatids of each chromosome are held together at a region called the *centromere*.

The chromosomes become aligned along one plane, the *equator*, which usually lies somewhere near the centre of the cell. Then the two chromatids of each chromosome migrate in opposite directions. The result is two complete, and identical, sets of chromosomes, around which nuclear membranes form. New cell formation obviously requires the production of a new membrane to separate the two nuclei. The distribution of the various organelles throughout the cell prior to division usually ensures that some of each type will be found within each of the two new cells. It is not only diploid nuclei and cells that can divide mitotically, but haploid nuclei and cells as well. The vital point is that a mitotic division gives rise to two daughter nuclei, or cells, that are genetically identical, and of course the number of chromosomes is the same as in the mother nucleus.

The rate of cell division varies during the life cycle of an organism, and between different types of cell. During an organism's growth and development, cells may be dividing almost continuously, but the pace of cell division tends to slow with age. Some cell types, once they have differentiated, undergo no further division.

Meiosis

Meiosis involves two successive nuclear divisions. In the first division, the chromosome number is halved (so only diploid nuclei can divide meiotically). In other words, haploid nuclei are produced, but each of the two new nuclei contains one of each type of chromosome and, where it applies, one type of sex chromosome.

The necessity for such a *reduction* division during the life cycle of sexually reproducing organisms should be quite obvious. An individual that is produced by sexual means begins life as a single cell, known as a *zygote*, which carries the diploid number of chromosomes. The zygote is the result of the union of special cells, or more strictly, special nuclei, known as *gametes*. (Eggs and sperm are female and male gametes respectively.) Now if two diploid nuclei were involved in such a union the number of chromosomes in the zygote would double, and each successive union would result in a doubling of the chromosome number. A reduction division is therefore essential. Then, when two gametes unite, the diploid number is restored. Furthermore, each gamete must have one or other 'version' of each type of chromosome, otherwise all the information necessary for a new individual would not be present and there would not be homologous pairs of chromosomes when gametes unite.

The behaviour of chromosomes during meiosis is shown in Fig. 3.3. As in mitosis, the DNA on each chromosome is copied and the nuclear membrane disappears. Each chromosome is now made up of two chromatids. Homologous chromosomes then come together (which is not the case in mitosis), a process called *synapsis*. Each chromosome 'unit', known as a *bivalent* or *tetrad*, therefore comprises four chromatids. Bivalents align along one plane as in mitosis.

Homologous chromosomes (each comprising two chromatids) separate and migrate in opposite directions. So each of the two new chromosome sets comprises just one 'version' of each chromosome, although in duplicate form. Also, and very importantly, the different 'versions' of the same chromosome type can migrate either way; it depends on how they line up initially. Consider a diploid cell with just two types of chromosome; the two homologues of one type being represented by A and a, the two homologues of the other type by B and b. As bivalents

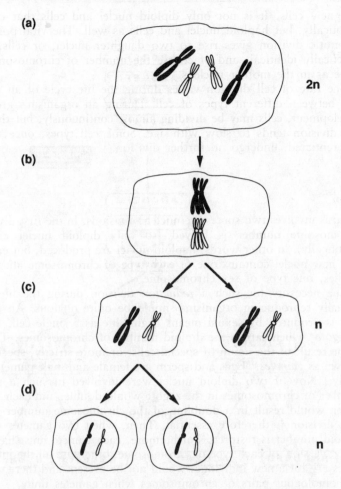

Fig. 3.3 The behaviour of chromosomes during meiosis. Two separate divisions are involved. Following duplication (a) homologous chromosomes come together (synapse), in which form they line up (b). The separation of chromosomes gives rise to two haploid (n) nuclei (c) containing one of each type of chromosome, in duplicate form. The new nuclei are therefore not genetically identical. During the second division, the duplicated chromosomes separate so that each new nucleus has one copy of genetic material. Both cells from each division do not necessarily survive.

behave independently, the two haploid cells resulting from the first meiotic division could carry AB and ab, *or* Ab and aB. (Actually, each 'version' is in duplicate form at this stage, having previously doubled.)

At the onset of the second meiotic division, the chromosomes again become aligned in a single plane. The chromatids then come apart at the centromeres and move in opposite directions. They are now referred to as chromosomes and each set becomes surrounded by a nuclear

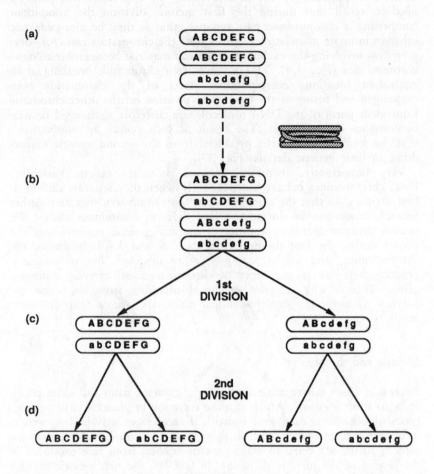

Fig. 3.4 The principle of crossing-over during meiosis. Different letters denote different genes; upper and lower case letters denote different alleles. (a) A pair of duplicated, homologous chromosomes lie together prior to the first meiotic division. Genetic material is exchanged between them during crossing-over (small diagram). New combinations of alleles are apparent after the first division (c). After the second division the four nuclei all have different allele combinations (d). (For the female of many types of organism only one nucleus may survive each division, but reassortment of chromosomes and recombination of alleles still occur during meiosis.)

membrane. The new nuclei are of course still haploid; however, there is now only one copy of the genetic material whereas prior to the second division the genetic material was in duplicate form.

So what purpose is served by the second division? After all, haploid nuclei are produced by the first division. To appreciate this point we need to recall that during the first meiotic division the homologues comprising a chromosome pair synapse, that is they lie alongside each other in intimate association. At this time the chromatids can *cross-over*, a process involving the exchange of genetic material between homologous chromosomes (Fig. 3.4). This involves two chromatids breaking at the equivalent location, corresponding parts of the chromatids being exchanged and fusing at the equivalent position on the other chromatid. Equivalent parts of the DNA molecule are therefore exchanged between homologous chromosomes. The result of such *genetic recombination* is that the four haploid nuclei produced from the second meiotic division differ in their genetic details (Fig. 3.4).

Very importantly, therefore, meiosis promotes genetic variability. First, chromosomes behave independently when they separate during the first division, so that the same 'versions' do not always migrate together. Second, crossing over during the first division recombines alleles. The second division therefore capitalizes on any genetic reassortment that occurs during the first division. In Figs. 3.3 and 3.4 only one or two chromosomes, and just a few genes, are depicted, but of course in practice there are several types of chromosome, all carrying numerous genes. This is why sexually produced offspring from the same two parents are never genetically identical unless the zygote splits following fertilization.

Meiosis and the life cycle

Except in cases where male and female gametes from the same parent unite to form a zygote, which is quite common in plants and some other types of organism, each new individual receives exactly half its genetic information from each parent. In most types of animal, meiosis gives rise directly to the sex cells. In males, sperm develop from cells produced by the second of the meiotic divisions. In females, the first meiotic division gives rise to two cells of very different size. The larger of these again divides to produce unequally-sized cells, the larger of which develops into the egg.

Other organisms behave differently. It is a characteristic of fungi that the first division of the zygote nucleus is a reduction division. So in the case of these organisms the only diploid nucleus is that of the zygote, all the rest are haploid.

With plants the situation is somewhat different again. Recall the point made earlier that the conspicuous 'bodies' of moss plants are made up of haploid cells; so clearly a meiotic division does not give rise directly to sex cells. Rather, successive mitotic divisions of haploid cells occur after meiosis. It is on the haploid generation that the sex cells are eventually produced. If sperm and egg unite, a new phase of the moss life cycle begins, and of course it is diploid. The diploid generation of mosses is small and short-lived, but on it special cells divide meiotically to produce

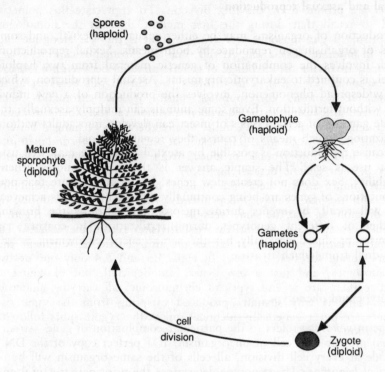

Fig. 3.5 The principle of alternation of generations in plants. The example used is a fern,
a non-flowering plant in which the sporophyte (diploid) generation is dominant
and the gametophyte (haploid) generation is inconspicuous and independent.

the new haploid generation. In other words there are two generations in
the life cycle of a moss, one is diploid, the other is haploid.

There are in fact two distinct generations in the life cycle of practically
all plants, a phenomenon referred to appropriately as the *alternation of
generations*. In some types of plant the two generations exist as independent entities, but in others they remain closely associated. The diploid
phase of the life cycle is called the *sporophyte* (generation) while the
haploid phase is called the *gametophyte* (generation). Mosses are
somewhat unusual among plants in that the gametophyte is the dominant
generation. For the 'seed' plants it is the sporophyte that is dominant; the
gametophyte is tiny and short-lived and its development takes place on
the sporophyte generation. For the flowering plants this is within the
floral organs. The principle of alternating generations in ferns, a type of
seedless plant in which the sporophyte is dominant, is depicted in Fig.
3.5.

Sexual and asexual reproduction

Reproduction of organisms may be either sexual or asexual, and some kinds of organism can reproduce by both means. Sexual reproduction, which involves the combination of genetic material from two haploid nuclei, is confined to eukaryotic organisms. Asexual reproduction, which is a widespread phenomenon, involves the production of a new individual without fertilization. Even some animals can multiply asexually; the female gametes of certain types of insect can develop into adults without fertilization, which means of course they remain haploid.

Because reproduction is possible by asexual means we may well ask, 'what use is sex?' The simple answer is that sex promotes genetic variability. Sex does not create new genes, but it does ensure that new permutations of genes are being continually thrown up. This is achieved, you will recall, by events during meiosis, and also by the bringing together of different genotypes during fertilization. In contrast, an organism produced asexually has the same genetic endowment as the individual from which it arose.

Genotype and phenotype

The term *genotype* refers to the particular combination of genes (strictly alleles) present in a cell or an organism. If a perfect copy of the DNA is made at every cell division, all cells of the same organism will be of identical genotype. The genotype determines the principal structural and functional features of an organism. However, an organism's environment has a modifying effect on growth and development; in other words the environment determines the extent to which the genetic potential is realized. Oak tree seedlings of identical genotype grown in different light environments will show quantitative differences in their morphology, although both are unmistakably oak trees. The environmental influence commences at conception: the offspring of animals that are maintained on a low plane of nutrition during pregnancy may never fully recover from that early check to growth and development.

So the genotype provides the ground plan for growth and development while the environment determines exactly how the genotype is expressed. Some features are genetically fixed, or stable, and not subject to environmental modification. Other features though are more susceptible to modification and alteration by the environment. Organisms vary also in the degree to which their morphology can be modified by the environment. As a group, plants are particularly susceptible to modification by their environment; they are said to exhibit *plasticity*.

The form and the functional characteristics of an organism collectively constitute its *phenotype*. The phenotype is determined by the interaction between the genotype and environment. As just implied, measurements made on two individuals of the same kind in their natural environment will not be very informative about their respective genotypes, except at a trivial and obvious level. We cannot be sure that phenotypic differences are determined genetically, and not environmentally. However, if we compare phenotypic characteristics of two organisms of the same kind raised in the same environment we may be able to deduce something about their genetic differences. We might find they grow at different rates, or they differ slightly in morphology. In the next chapter we look at the significance of phenotypic and genotypic variation for evolution.

4 Evolution

In an everyday sense, evolution means gradual change; in a biological context, evolution means a directional change in the frequency of genes in a population over time. (Recall that in Chapter 1 we defined a population as a collection of individuals of the same type inhabiting a prescribed area.) In other words, evolution is characterized by a change in the composition of a population's *gene pool* over time. Natural selection is a process by which evolution occurs in nature. It is important not to confuse evolution and natural selection; the terms cannot be used interchangeably.

The one name above all others that we associate with the theory of evolution by natural selection is Charles Darwin (1809–82). However, Darwin was not the first person to believe in evolution; his special contributions were, first, to provide overwhelming evidence that evolution occurred and, second, to propose a mechanism, natural selection, by which it could do so. In this chapter we will first summarize the essential principles of natural selection and then briefly return to Darwin because of his enormous contribution.

Variation and natural selection

As a starting point we shall consider a population of organisms. The performance of individuals in a population always varies: for example, some individuals will grow more quickly than others, some will be more resistant to disease, withstand periods of adverse weather better and so on. Most importantly, individuals vary also in their reproductive success, with the result that the genes of some individuals are better represented than the genes of other individuals in future gene pools.

From our discussion in Chapter 3, it will be clear that differences between individuals are partly genetic and partly due to environmental factors. So the differences in performance that we can observe and measure are phenotypic differences. However, because an organism's genes have a strong influence on its performance, it is reasonable to assume that *on average* the phenotypically 'superior' individuals are also genetically 'superior' in that particular situation. In other words, differences in reproductive success between phenotypically different individuals will lead to a shift in the composition of a population's gene pool. Differential reproductive success between genotypes leads to a change in gene frequency from one generation to another.

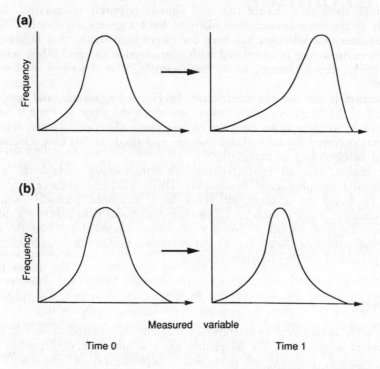

Fig. 4.1 (a) Directional selection. The frequency distribution of the measured trait (e.g. height, length, weight) has changed with time. Selection is evidently for larger individuals. (b) Stabilizing selection. The mean and mode of the frequency distribution have not changed with time. Selection is for individuals which were most abundant at Time 0. Although it is phenotypic characters that are measured, it is assumed they are correlated with genotype.

Fitness is the term used to refer to the relative success of genotypes in passing on genes to successive generations; the fittest individuals being those that contribute most genes to future gene pools. So, in an evolutionary context, fitness has a precise meaning which is somewhat different from its common usage. Any genetically determined characteristic can contribute to fitness. It could be the ability to grow faster, to run more quickly, to withstand drought better, or it could be some behavioural characteristic. However, a trait can be said to contribute to fitness only if its possession increases the probability that genes will be passed on to future generations.

A change in the characteristics of individuals comprising a population through successive generations can be represented diagrammatically using frequency distributions as in Fig. 4.1. Here we assume a characteristic that varies continuously, body dimensions and weight for example, is measured on all individuals at Time 0 and on their descendants at Time 1. In Fig. 4.1a, population characteristics have clearly shifted between Time 0 and Time 1; mean and mode values have moved to the right.

Although there is a chance that this change is purely phenotypic, due simply to the environment, it is likely to have a genetic component. We will assume that selection has been for larger individuals, that largeness in this environment is correlated with reproductive success. When selection brings about change, as in this example, it is described as *directional*.

Selection is not always directional. In Fig. 4.1b, means (and modal values) of the frequency distributions are much the same at Time 0 and at Time 1, so there is no evidence of directional selection. Rather, selection has occurred for individuals that are mid-sized; in this case selection would be described as *stabilizing*.

Of course, not all characteristics are continuously variable as are weight and morphological dimensions. Differential reproductive success may be based on a small difference in a metabolic pathway. Some individuals in a population could be more successful than others because of their capacity to break down a poisonous chemical released by an invading pathogen. Selection would therefore favour those individuals that possessed the appropriate enzyme.

Adaptations

Genetically determined characteristics that enhance reproductive success are called *adaptations*. We frequently say that one type of organism is well adapted, pointing to particular morphological or physiological characteristics that appear to equip it well for the environment in which it lives. Strictly though, characteristics should be described as adaptive only if their possession demonstrably enhances reproductive success.

Study of an organism may suggest aspects of its form, function or behaviour that appear to enhance its survival and reproductive success in a particular environment. When we use the term 'environment' here, we mean all the various factors that impinge on an organism throughout its life; the physical environment of course (temperature, moisture, nutrients, etc.), but also the biotic environment, that is, other organisms that may be predatory, cause disease or compete for resources. We can regard the environment as a type of filter, or series of filters, selecting some genotypes but discarding others.

However, it does not follow that all organisms are perfectly adapted to their existing environments. What we observe today are really the net result of past selection pressures. In this context it is important not to forget that the environment is not constant over time. Genotypes that are well adapted to one set of conditions may be poorly adapted if conditions change. Some organisms though, particularly some plants, can respond to a certain amount of change in their environment by altering some aspect of their physiology or development; they are said to exhibit *phenotypic flexibility*.

From careful studies of a population it may become apparent which aspects of the environment are important from an evolutionary point of view; in other words, which environmental factors are providing high selection pressure and at what stages of the life cycle they are most critical. It might be fire, salinity or drought, or it could be another type of organism that is important. However, the environment of an organism is multifaceted and complex; weighting the importance of various environmental factors is usually problematic in practice. Thus the form, function and behaviour of organisms usually represent a compromise response to the total environment. It is normally only in extreme environments that a single factor can be identified as being of overriding significance.

Sometimes, similar characteristics are evident in two or more unrelated types of organism, suggesting they have been subject to similar selective forces and have responded (in an evolutionary sense) in somewhat similar ways. This phenomenon is known as *convergent evolution*. *Divergent evolution*, as its name suggests, means that characteristics of two populations are becoming increasingly different. *Parallel evolution* is said to occur when two types of organism which share a common ancestor but subsequently diverged, appear to be evolving in similar ways.

Selection, variation and reproductive potential

For selection to occur there must be *genetic variation*. This is simply because variation is the only raw material available on which natural selection can operate. Natural selection cannot create new genes, it can only work on what is available in the gene pool. Central also to natural selection is the fact that all types of organism potentially leave more offspring than is necessary simply to maintain the population at its existing level of abundance. The potential number of offspring produced by one individual in a single generation can be truly enormous, as is the case with eggs produced by fish and insects and the seeds produced by flowering plants. Even populations of large mammals, with their characteristically low reproductive output per year, have a high potential for population increase. It is clear though that potentials for population increase are not realized; in general populations remain more or less stable over quite long periods of time, although many fluctuate quite violently over comparatively short periods. Clearly then, either most offspring die before reaching reproductive age or else they fail to reproduce.

The importance of the potential for population growth in the context of a discussion of evolution is that, on average, those individuals genetically better equipped for their environment are more likely to live to reproductive age, and hence leave their genes to succeeding generations, than genotypes that are less well equipped.

Examples of natural selection

There are a number of convincing examples of evolution by natural selection that we could cite. Frequently, human agencies are involved because of our unrivalled capacity to change the nature of selective forces very quickly. Probably the most quoted example is that of a moth, commonly known as the peppered moth, in England during the nineteenth century. Prior to the industrial revolution, trees were typically covered by lichen of generally light colour. However, the deposition of air-borne pollutants resulted in the loss of the lichen cover in and around industrial areas, thus exposing the dark-coloured bark beneath. When trees were covered by lichen, peppered moths were predominantly light in colour, and therefore well camouflaged from predatory birds. Dark-coloured forms of the moth were rare. As the trees lost their lichen cover, light-coloured moths became more conspicuous than dark-winged forms and therefore more vulnerable to predation. An increase in dark, or melanic, forms occurred at the expense of light-winged forms. Wing colour is a genetically determined characteristic. Thus, predatory birds selected for genes specifying dark-coloured wings in these peppered moth populations.

Background colour has been shown experimentally to have an effect on the rate of predation within peppered moth populations. However, if predation had not been a major cause of death, the loss of lichen might not have had any effect on the frequency of genes specifying wing colour in the moth population. Of course it should not be assumed that predation is the major selective force in every population; as we pointed out earlier, populations are subjected to a variety of potentially important selective agents in their environments.

Further evidence for the efficacy of appropriate selection pressures to change gene frequency is provided by the numerous cases of evolution of pesticide resistance documented for a variety of pest organisms, and similarly the development of resistance to antibiotics within bacterial populations.

The examples cited above demonstrate the rate at which evolutionary change can occur in response to powerful selective pressures; as a general rule, the greater the selective force on a particular gene pool, the greater the pace of change in gene frequency. Not surprisingly, the pace of change is potentially greater the higher the reproductive rate and the shorter the time between generations.

Artificial selection

Artificial selection is the deliberate selection of individuals for breeding purposes. It is of course phenotypes that are normally selected, the assumption being that phenotypic differences among individuals will have some genetic component. Selection and breeding of some organisms was carried out for thousands of years in complete ignorance of genes and

genetics. For agricultural purposes, desirable traits frequently include
high yields and resistance to disease. Artificial selection has provided us
with modern varieties of crops and livestock, ornamental plants and
domestic pets.

The products of artificial selection convincingly demonstrate the
considerable genetic variability that usually exists within the gene pool of
a particular type of organism, and therefore the potential for change
unrealized under natural conditions. Natural selection did not produce
dogs as diverse as the beagle, poodle and spaniel, or modern wheat and
corn varieties: the potential was always there, but it has been realized
only by deliberate selection and intensive breeding. Dog breeders and
wheat breeders are not necessarily concerned with features that enhance
survival and reproductive success in a natural environment. The
environments in which wheat is grown and poodles are raised are greatly
modified, so attention is focussed on other characteristics of the
organism. Furthermore, plant and animal breeders can utilize the full
range of compatible genetic variation that is available, whereas under
natural conditions genetic exchange is normally a quite local affair.

Genetic engineering

We are living at a time of rapid progress in molecular biology. Of
concern to us here are techniques which permit bits of DNA specifying
particular characteristics in one organism to be located, isolated and then
transferred to another organism in which they do not naturally occur.
Techniques of this sort, known as *recombinant DNA technology*, are
central to developments in biotechnology. Potentially, these techniques
enable the available gene pool to be expanded much more widely than
is possible with conventional breeding techniques. Because of the universal nature of the genetic code, genes from one type of organism can be
transferred to other organisms of very different kinds. Transferring genes
that specify certain characteristics between different types of organism
may have a variety of useful applications. It may, for example, permit
frost tolerance or disease resistance to be enhanced in organisms that are
presently vulnerable. However, it should not be assumed that anything
is possible with recombinant DNA technology; there are a number of
biological constraints. Nevertheless, such technology does permit an
extraordinary degree of biological manipulation at a fundamental level.
Not surprisingly, however, progress in genetic engineering has brought
with it problems, both practical and ethical. How, for example, can we
predict what the consequences might be of releasing a new type of
bacterium into the environment? It could possibly prove to be
pathogenic, causing disease outbreaks among humans or crops or
livestock.

Mutations

We have stressed that genetic variation is the only raw material on which natural selection can operate, and on which artificial selection can draw to produce genotypes which meet specific objectives. However variable, the gene pool sets a limit to the amount of genetic change that can occur. Neither natural nor artificial selection can add to existing variation by bringing into existence new genes, only new combinations of genes.

A permanent change in genetic material, that is a change in the DNA of a cell, is referred to as a *mutation*. A mutation involves a change in the sequence of nucleotide bases along a DNA molecule and hence has implications for the ordering of amino acids and the type of proteins produced. Mutations occur very occasionally under natural conditions, and can be induced by external agents such as ionizing radiation and certain chemicals. A change in nucleotide sequence, through its effects on protein synthesis, has the potential to alter some aspect of form or function in the cell carrying the 'new' gene. (The mutated gene might not be 'read' at all of course.) Mutations that are incorporated into the gene pool give rise to new genes and, importantly, compensate for those lost from the population because of the failure of some individuals to reproduce.

A mutation may be advantageous in the sense that it increases an organism's chances of passing on its genes to successive generations; or it may be disadvantageous. In fact, most are soon lost from the gene pool. The important point is that genes that have just arisen by mutation are not special in any way; they should just be considered part of the gene pool and therefore part of the raw material for natural selection.

Alternative forms of the same gene, i.e. alleles, arise because of mutations, and all organisms are, by definition, the products of successful mutations in the past. However, mutations do not bring about great changes in genotype within one generation. If large numbers of genes were involved, normal cellular processes would be so disrupted that the cells would cease to function. Furthermore, under natural conditions, the rate at which mutations occur is very low. Nevertheless, mutations are the only source of brand new genetic material and are therefore central to any explanation of the diversity of life.

Evolution, complexity and diversity

We have characterized natural selection as a process operating on existing genetic variation in such a way as to bring about a progressive change in gene frequencies within a population. Genes are lost from the gene pool through failure of individuals to reproduce, and new genes are very occasionally introduced by mutation. The gradual nature of evolution has been emphasized, notwithstanding the fact that changes in gene frequency can sometimes be observed in nature over relatively short intervals of time. At this stage we may well ask whether known evolutionary processes can alone explain the enormous diversity of life on our planet, all the complexities of structure, function and behaviour displayed by organisms, and the history of life as revealed by the fossil record. The conventional answer is 'yes'. And the reasons that the answer can be given so confidently are that a proven mechanism for evolution exists, the time-scales we are dealing with are so very long, and anyway the many unifying features of living cells argue strongly for a common ancestry.

Darwin and *The Origin of Species*

Charles Darwin's great synthesis, *The Origin of Species*, proposing natural selection as the mechanism for evolution (although he rarely used the term 'evolution') was published in 1859. Darwin is of course widely remembered for his time on HMS Beagle, which sailed from Plymouth in 1831 on an expeditionary voyage to the southern hemisphere, and particularly for his observations on the Galapagos Islands in the eastern Pacific. However scholars of Darwin are quick to refute the suggestion that he hit upon natural selection in a sudden stroke of inspiration, and particularly that any such inspiration occurred on the Galapagos Islands. However it would be true to say that Darwin's observations of geographical variation among organisms on these islands contributed towards the development of his ideas on biological change. In addition, on returning to Britain, Darwin had the opportunity to discuss his observations and the many specimens he collected with the experts of the day.

There were other influences as well. Darwin took with him on the voyage the first volume of Charles Lyell's classic work, *Principles of Geology*, which had just been published (the other two volumes were sent on to him while abroad). Seemingly, Lyell's advocacy of slow and gradual change in the natural world, and also the importance of understanding contemporary processes for explaining past changes, provided Darwin with guiding principles.

Important also to the development of Darwin's theory of natural selection were the ideas contained in Thomas Malthus's *Essay on the Principles of Population* which had been published in 1798. Malthus's principal thesis was that numbers of individuals in human populations tend to increase in a geometric progression from generation to generation (i.e. population number multiplies by a constant). However, the amount of food produced by human societies increases only arithmetically (i.e. the difference in amounts of food produced during successive intervals remains constant). Therefore, Malthus argued, food resources ultimately set a limit to population size. The potential for an increase in numbers is a characteristic of all plants and animals, so for Darwin, an inevitable outcome of a geometric rise in numbers was a 'struggle for existence' between individuals. So Malthus's observation concerning the reproductive potential of organisms is a key feature of the theory of natural selection which requires differential reproduction in a potentially expanding population according to how well organisms are adapted to their environment. In addition, Darwin appreciated the power of artificial selection and breeding to bring about organic change, and indeed devoted the first chapter of *The Origin* to this subject.

The theory of natural selection is associated also with a contemporary of Darwin, the British naturalist Alfred Russel Wallace. Wallace, who was working in the Malay Archipelago (modern Indonesia) in the 1850s, wrote to Darwin describing his ideas on natural selection; these were essentially similar to Darwin's own, which although by now well developed, had yet to be published. In 1858, papers containing the ideas of the two men were read, in their absence, to a small audience at the Linnean Society in London and appear to have had little impact. But in 1859, Darwin, spurred by the knowledge that he was not the only person to have discovered natural selection, finally published *The Origin of Species*, unquestionably one of the most influential books ever written. It quickly generated fierce debate among scientists, philosophers and church people, and a wider public as well, and it was to revolutionize scientific thought.

5 Energy and life

Energy is defined as the capacity to do work; living organisms are working entities and as such expend energy continuously. The more work being done, the greater the energy expended. So the rate of energy expenditure of a seed lying dormant in the soil is only a fraction of that expended by the same seed when it is germinating; similarly, the rate at which an animal expends energy depends on its level of activity.

The energetics of life can be studied at various levels of organization; at molecular and subcellular level, at the levels of the cell, the individual organism, and the community of organisms. We can also consider the energy relations of the entire biosphere. In Chapter 9 we shall consider energy flow through ecological communities, but first it is necessary to introduce some basic principles concerning energy and living organisms.

Introducing energy

Whenever work is being done, energy in some form or other is involved. Many of the different forms of energy are part of everyday vocabulary; electrical energy, solar energy, nuclear energy and chemical energy: heat is another form of energy. For work to be done there must be movement of some sort, so a distinction is made between *potential* energy and *kinetic* energy, or the energy of motion. A boulder resting at the top of a hill has potential energy by virtue of its position, but only if it is rolled or dropped can that potential for work be realized.

Within cells and organisms, usable energy is present in chemical form. Energy may be stored for relatively short periods in small molecules such as those of glucose, but for longer periods energy is stored in larger, *macromolecules* such as starch and the triglycerides. This potential energy is made available by the rearrangement of atoms or by the breakdown of the molecules.

Chemical reactions that involve the release of energy are termed *exergonic*; the energy content (energy state) of the product(s) is less than the starting material. Other chemical reactions, termed *endergonic* require an input of energy in order to proceed. In general, whenever organic molecules are broken down, energy is released, but the synthesis of organic molecules requires an input of energy. The standard unit for the measurement of energy values is now the *joule*, although in the older literature the *calorie* was frequently used, and is often still used in connection with human diets.

One of the basic rules of science is that energy may be transformed from one form to another, from chemical energy to heat for example, but it can neither be created nor destroyed. So, if molecules of glucose are broken down to carbon dioxide and water, the energy formerly contained within those glucose molecules must be around somewhere in the universe, and in one form or another. Here we need to invoke a second rule, which relates to the quality of energy: when energy is used, it is converted to a more dispersed, less useful, form; in other words it is degraded. Important here is the fact that when chemical energy is used within a cell, heat is released. These two rules, one pertaining to energy quantity and one to energy quality, constitute in essence the first two laws of thermodynamics, and it is helpful to keep them in mind.

Autotrophic and heterotrophic organisms

The chemical reactions involved in the processing of energy by an organism collectively constitute its *energy metabolism*. Our objective here is to understand the barest essentials of this subject because it is central to the functioning of the biosphere.

We can assign practically all organisms to one of two groups on the basis of how they meet their energy needs. One group, the *autotrophs* harness an external, nonbiotic, energy source to manufacture energy-rich organic molecules from carbon dioxide. The other group, the *hetero-trophs*, lack this capacity and are thus dependent on a dietary supply of preformed organic molecules. These molecules are then broken down to supply the energy needs of the organism. Heterotrophic organisms, such as animals and fungi, consume living material or organic remains; they thus depend on other forms of life. Autotrophic organisms, in contrast, are independent of other organisms as far as their energy needs are concerned. A few microorganisms can switch metabolism between auto-trophic and heterotrophic modes, and there are examples of mutualistic association between autotroph and heterotroph, but these are comparatively rare.

There are two modes of autotrophic nutrition, namely *photosynthesis* and *chemosynthesis*. Photosynthesis involves the 'trapping' of light energy and its conversion to chemical form. Chemosynthesis, which is confined to certain groups of bacteria, involves the utilization of energy of certain inorganic chemical entities that are scavenged from the environment. We shall look at the processes of energy metabolism a little more closely.

Photosynthesis

The conversion of solar energy to chemical energy in photosynthesis underpins all life on the planet. The most familiar photosynthetic organisms are of course the green plants, but the unicellular algae of aquatic ecosystems, and also some groups of bacteria, including the cyanobacteria (formerly called blue-green algae) are photosynthetic. The green colour of photosynthetic organisms is due to the pigment *chlorophyll*, molecules of which are intimately involved in the photosynthetic process.

During photosynthesis, hydrogen combines with carbon dioxide to produce simple organic molecules. For all but a few types of photosynthetic organism, water is the source of hydrogen, and oxygen, derived from the split water molecules, is released. The process may be expressed simply as:

$$2H_2O + CO_2 \rightarrow (CH_2O) + O_2 + H_2O$$

Here, (CH_2O) represents a simple carbohydrate (often a molecule of glucose is represented in such expressions).

The exceptions are certain groups of photosynthetic bacteria that operate in the absence of oxygen, i.e. in *anaerobic*, or *anoxic*, environments. Most of these organisms use hydrogen sulphide (H_2S) as the source of hydrogen, thus:

$$2H_2S + CO_2 \rightarrow (CH_2O) + 2S + H_2O$$

Notice that in this case it is sulphur, not oxygen, that is released. It is believed that this was the first type of photosynthesis to appear on the planet, at a time when there was no free oxygen present.

It is important to be aware that these shorthand expressions for photosynthesis show only the raw materials and the end products. It is not a one-step process, however, but involves a series of biochemical and biophysical transformations leading to the formation of fairly simple organic compounds. We should be in no doubt as to the importance of this process in the biosphere; all life on Earth, directly or indirectly, depends on photosynthesis. It is because of its central importance that we devote most of a later chapter to photosynthesis.

Despite fundamental differences, autotrophic and heterotrophic organisms share certain features of their energy metabolism. Note that both types of organism disassemble energy-rich organic molecules in order to provide energy for cellular processes; photosynthetic organisms do not stop 'working' at night when no light energy is available, and anyway not all parts of the plant are photosynthetic. So within the cells of autotrophic and heterotrophic organisms energy is made available in a similar fashion: the key difference between them is that whereas autotrophs can manufacture organic molecules from carbon dioxide using an external source of energy, heterotrophs require preformed organic substances.

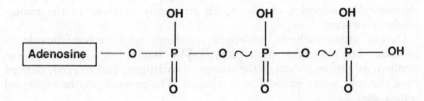

Fig. 5.1 Schematic representation of an ATP molecule showing its three phosphate groups and high energy bonds ($\sim$). The hydrolysis of ATP is a highly exergonic reaction.

The importance of ATP

We cannot proceed very far in a discussion of energy and life before meeting a comparatively simple compound called adenosine triphosphate, or ATP. ATP is found in every living cell, regardless of type and regardless of organism, and is nearly always involved when work is being done within a cell. For this reason ATP is sometimes referred to as life's 'universal energy currency'.

A molecule of ATP is depicted in Fig. 5.1, highlighting the three phosphate groups that are a key feature of the molecule. The bonds linking the outermost and middle phosphates, and the middle and innermost phosphates, are known as 'high-energy' bonds, and are conventionally depicted by 'squiggles'. As its title suggests, the formation of a high-energy bond requires a substantial input of energy. However, hydrolysis of such a bond, in a coupled reaction, provides energy for endergonic reactions. Such endergonic reactions include the contraction of muscle fibres, transmission of nerve impulses, pumping of ions across membranes and the synthesis of organic molecules. In other words the energy associated with ATP molecules is harnessed (although not with one hundred per cent efficiency) for useful work during coupled chemical reactions.

When a molecule of ATP is hydrolysed, adenosine diphosphate (ADP), and inorganic phosphate (P_i) are produced, thus:

$$ATP + H_2O \rightleftharpoons ADP' + P_i + energy$$

As we see, the reaction is reversible; ADP can link up with a phosphate group to form ATP. The formation of ATP requires an input of energy of course. So where does this energy come from? This takes us back to the distinction between autotrophic and heterotrophic organisms. Actively photosynthesizing cells generate ATP using light energy

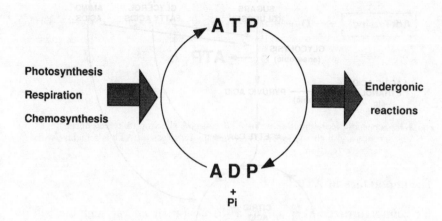

Fig. 5.2 ATP generation. The processes listed on the left-hand side result in the
         generation of ATP from ADP and inorganic phosphate (P_i). Photosynthetic
         organisms use light energy while chemosynthetic organisms oxidize simple
         inorganic chemicals. Autotrophic organisms use ATP to power the manufacture
         of organic chemicals from carbon dioxide. Heterotrophic organisms need
         preformed organic molecules which they break down to generate ATP.
         Autotrophic organisms also break down the organic molecules they manufacture
         for ATP production. The conversion of ATP to ADP is coupled to the many
         energy-demanding metabolic activities in the cell.

'harvested' by the photosynthetic apparatus. Chemosynthetic organisms
use certain simple chemical entities as an energy source. Heterotrophic
organisms, however, depend entirely on the energy present in the
preformed organic molecules that they consume. (But, as indicated
earlier, the organic molecules manufactured during photosynthesis may
be disassembled later by the same organism to generate ATP, just as
organic molecules are in the cells of heterotrophic organisms.) An outline
scheme showing these various routes to ATP generation and ATP
hydrolysis is presented in Fig. 5.2.

Energy metabolism

Energy metabolism involves the disassembly of organic molecules within

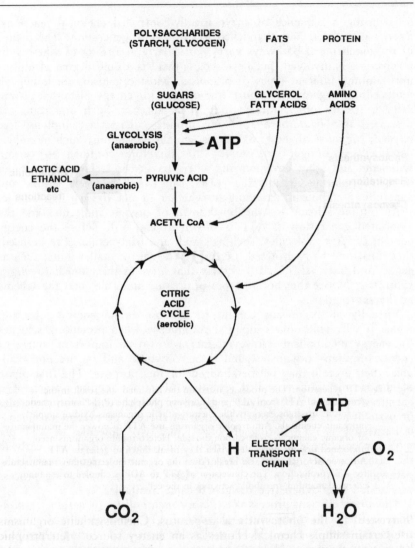

Fig. 5.3 Energy metabolism. An outline scheme for the breakdown of large organic molecules and the generation of ATP within the cell. Both glycolysis and the citric acid cycle comprise several separate chemical reactions. Carbon dioxide and hydrogen are released during reactions of the citric acid cycle; the hydrogens pass along the electron transport chain and finally combine with oxygen to form water. The chemical products of respiration are thus carbon dioxide and water. Fermentation, an anaerobic process, leads to the generation of substances such as lactic acid and ethanol according to the kind of organism. Oxidative energy metabolism yields far more energy, as ATP, than anaerobic energy metabolism.

cells during a sequence of enzymatically-controlled chemical reactions. Energy is released and channelled towards ATP generation. The details of the biochemical pathways vary, particularly according to whether or not oxygen is involved, but nevertheless there is a high degree of uniformity among different types of organism. Most organisms, including all multicellular organisms, require free oxygen for energy metabolism, and can survive for only short periods in its absence. Such organisms are described as *obligate aerobes*. A few groups of bacteria cannot use free oxygen; for most of these, oxygen is actually poisonous. Such microbes, described as *obligate anaerobes*, are therefore confined to certain sediments and other environments where oxygen is lacking. Some types of microorganism, notably the yeasts, can switch between aerobic and anaerobic metabolism according to whether or not oxygen is available.

An outline scheme for the breakdown of organic molecules and the associated generation of ATP is shown in Fig. 5.3. Before the energy content of large molecules, such as starch and triglycerides, can be used, they must first be hydrolysed, i.e. broken down to smaller units, such as sugars and fatty acids, and it is these that serve as the immediate *energy substrates*. Notice that not all types of organic molecule enter the scheme at the same point.

Virtually all organisms can utilize glucose as an energy substrate, which is why textbooks usually show glucose when presenting schemes for energy metabolism. However, fats are also an important source of energy for many organisms, and if carbohydrate and fat are not available, then protein may be metabolized for this purpose. The first phase of glucose metabolism, called *glycolysis*, does not require oxygen. Ten chemical reactions are involved in glycolysis, which 'ends' with molecules of pyruvic acid. Only a small fraction of the energy present in glucose is harnessed as ATP during glycolysis; just two molecules of ATP per molecule of glucose in fact. Glycolysis must have evolved very early in the history of life; it can be regarded as a fairly primitive metabolic pathway on to which a much more efficient energy-harnessing process was added later when free oxygen became available.

This more efficient process is *oxidative metabolism*, or *aerobic respiration*, and it yields thirty-four molecules of ATP per molecule of glucose, compared with the two ATP molecules generated during glycolysis. Respiration involves a biochemical pathway known variously as the *citric acid cycle*, *tricarboxylic acid cycle* or the *Krebs cycle* (after the Nobel laureate Sir Hans Krebs whose work led to its discovery). The word 'cycle' is used because the molecules that enter this series of reactions are only partly disassembled; the molecules remaining are available to combine with new molecules which are fed into the cycle. The citric acid cycle commences with molecules of the rather formidably-named substance, acetyl-CoenzymeA. Acetyl-CoA is produced from pyruvic acid, but as we see from Fig. 5.3, it is also derived from the breakdown of some other substances. The citric acid cycle comprises a sequence of nine reactions, each under the control of a specific enzyme.

During the citric acid cycle, carbon dioxide is released. Hydrogen atoms are also removed. (A hydrogen atom consists of a single proton and a single electron, but no neutrons.) These hydrogen atoms are then passed between a series of specialized carrier molecules, along what is known as an *electron transport chain*. Substances that lose electrons, or hydrogens, are said to be *oxidized* (irrespective of whether oxygen is involved); conversely, substances that gain electrons, or hydrogens, are said to be *reduced*. So the hydrogens, removed from molecules at various points in the citric acid cycle, are involved in a series of oxidation-reduction reactions; the molecules donating the hydrogens are oxidized, those molecules accepting the hydrogens are reduced.

In a series of such oxidation-reduction reactions, hydrogen atoms are transferred along the electron transport chain. They are passed to membranes where the electrons are stripped away leaving 'naked' protons. A proton gradient, i.e. an electrical gradient, across the membrane, constitutes a 'proton pump' which powers the generation of ATP. The hydrogens are eventually picked up by oxygen atoms and water molecules are generated. The oxygen is described as the *terminal electron acceptor*. (Actually, a few kinds of bacteria use alternative inorganic groups, such as nitrate, as terminal electron acceptors in a process referred to as *anaerobic respiration*.)

In outline form, respiration can thus be represented as follows:

$$C_6H_{12}O_6 + 6O_2 \rightarrow 6CO_2 + 6H_2O + energy$$

(As we said earlier it is only a convention to show glucose because it is such a common respiratory substrate.)

Notice that the by-products of aerobic metabolism, carbon dioxide and water, are the chemical raw materials for photosynthesis. So the energy metabolism of cells is fundamental to the circulation of certain chemical elements in the biosphere, a topic we return to later.

The mechanism by which ATP is generated during photosynthesis is somewhat similar to that during oxidative metabolism. In fact it seems probable that oxidative metabolism evolved from photosynthesis.

In virtually all types of eukaryotic organism, oxidative energy metabolism occurs within special organelles called *mitochondria*. In aerobic prokaryotic cells, however, the same processes take place in the cellular cytoplasm.

Under anaerobic conditions, the first phase of glucose metabolism, i.e. glycolysis, is the same as when oxygen is present. However, the pyruvic acid generated from glycolysis does not enter the citric acid cycle (no such pathway 'exists' in obligate anaerobes). Instead, one or two chemical reactions occur yielding certain other substances. This process is called *fermentation* (although this term is frequently used to include the preceding glycolysis phase as well). It involves the addition of hydrogens (i.e. reduction) and is vital to the energetics of glycolysis under anaerobic conditions. The various end products of fermentation are characteristic of the type of organism. Animal cells that are deprived of oxygen produce a substance called lactic acid; plant cells typically produce ethanol, an alcohol, while the bacteria collectively yield a variety of fermentation products, including lactic acid and ethanol. The fermenting capacity of certain microbes is put to several good uses; to produce wine and beer, for example, in the case of yeasts supplied with sugar under anaerobic conditions. Carbon dioxide is also released as a by-product of yeast fermentation. As this gas is such an effective rising agent yeasts are extensively used in baking.

It is important to remember that oxidative metabolism is far more efficient than anaerobic energy metabolism in terms of the yield of ATP, and therefore in the amount of work that can be done from a given amount of the same energy substrate. As we discuss later this has important implications for ecological systems: under aerobic conditions, dead organic matter is broken down a lot more quickly than under anaerobic conditions.

Along with the chemical products of respiration and fermentation, heat is released of course. Non-biological analogies for the biological transformation of chemical energy to heat readily come to mind, as when coal is burned in a fire. When petroleum is burned in an internal combustion engine to provide work in the form of mechanical energy, there is a considerable temperature rise, requiring a cooling mechanism to prevent the engine from overheating.

The energy status of biological substances

Because chemical energy can be transformed to heat, the energy content of a biochemical substance can be determined by measuring the heat generated when a sample of known weight is combusted. The energy content of carbohydrate and protein are similar at about 16.0 and 17.0 kilojoules per gram respectively, but fats have over twice the energy content, about 38.0 kilojoules per gram (a piece of information essential for anyone on a diet). It is important to bear in mind that the total energy value of a substance as determined by its combustion is not necessarily the same as its biological energy value. Consider a packet of cornflakes. If we measured by combustion the energy status of an equal weight of packet and contents, the values would not be very different. This is because the packet is composed mainly of cellulose while the cornflakes themselves are largely starch, both of which are glucose-based polymers. However, the biological energy value of the packet is practically zero for humans, simply because we do not produce an enzyme that can hydrolyse cellulose to glucose, although we can hydrolyse starch. For a microbe that produces a cellulose-hydrolysing enzyme as well as a starch-hydrolysing enzyme the biological energy value of cellulose and starch could be quite similar.

6 *The diversity of life and the classification of organisms*

In previous chapters we have stressed unifying characteristics of life on Earth. At the same time though, it is abundantly clear that however many subcellular features and mechanisms are shared between organisms, a bewildering variety of shapes, sizes and life history patterns exists on our planet. Considering size alone, the range extends from tiny bacteria that are measured in micrometres to some whales which may be thirty metres or more in length and which weigh several tonnes.

There is clearly a need to organize this tremendous organismic diversity in some way, in other words to classify the many different types of organism. The science of classification, called *taxonomy* is an important part of an area of biological study known as *systematics* which deals with all aspects of relationships between organisms. In this chapter we introduce some basic principles of taxonomy and look at the structure of taxonomic schemes. We consider what is meant by a *species*, as it is species that are being classified, and also how new species arise.

Some basic principles

First, it is important to appreciate that classification schemes are entirely the product of human interpretation of biological variation. Organisms are classified according to their features, referred to as *characters*. Characters are used to distinguish one organism from another and for assigning individual organisms to particular taxonomic groups. The taxonomist decides which features are useful for classification purposes. Obviously, morphological characters are very important, but other characteristics used include information about the chromosomes, chemicals produced and metabolic pathways. Such features may be used to supplement morphological information or, as is common for some microorganisms, provide a basis for classification. Classification can be objective only if the variation exhibited by the chosen character is discontinuous. If variation is continuous, assigning individuals to particular groups is an arbitrary process.

The question then arises as to how characters are chosen, because different characters are likely to generate different classification schemes. As a general rule, characters which are *stable* are chosen; stable characters being features that are genetically determined and are not susceptible to environmental modification. A classification scheme for flowering plants that used the height of the stem, or number of leaves, would be of little use because these features are largely environmentally determined. On the other hand, floral structures are stable characters and provide a satisfactory basis for flowering plant classification.

Approaches to classification

Effectively, there are two approaches to modern scientific taxonomy. One, the *phenetic* approach, is based solely on the degree of resemblance between organisms. As indicated above, different characters are used in different major groups of organisms, and certain key characters are chosen within each. Classification schemes based on one or just a few arbitrarily chosen characters are termed *artificial*. *Natural* systems of classification are based on the simultaneous use of many characters, an approach known appropriately as *numerical taxonomy*. For types of organism that are difficult to classify in any other way, numerical taxonomy is particularly useful, but for most familiar organisms its use has been rather limited.

The other major approach to classification is based on evolutionary history, i.e. the *phylogeny*, of the organisms concerned. The *phylogenetic* (or *phyletic*) approach recognizes that groups of organisms shared a common ancestry at some time in the past, but have subsequently diverged. Phylogenetic relationships are usually represented by evolutionary 'trees', which indicate the time of divergence as well as the degree of resemblance between the organisms in the scheme. The phylogenetic approach to classification arose logically from an acceptance of evolution. However, its use is constrained by the extent to which the evolutionary history of the group of organisms under consideration is known. Such classification schemes usually require some informed speculation about the pattern of descent.

The kind of information likely to be useful in establishing evolutionary relationships, i.e. overall resemblance of living groups, is similar to that used to construct phenetic schemes. This is because phenetic schemes are likely to separate characters that evolved early from those that evolved later. So, phenetic and phylogenetic classification schemes can be quite similar, even though the former do not deliberately incorporate evolutionary information.

We should not forget that practicality is a major consideration when constructing taxonomic schemes. This is simply because the identification of organisms is one of the chief uses of classification schemes.

The species concept

The basic unit of classification in biological taxonomic schemes is the species; in other words, species are the objects that are being classified. We have used the term 'species' before in this book, but have not defined it. Now we need to give some thought to what the term means.

It would be convenient at this point to provide a short, unequivocal, definition of 'species', but in fact the term defies close definition. Of course, all members of a species should have much in common and be capable (where appropriate) of interbreeding, but such statements beg a number of questions and from them we cannot derive a definition. There may be considerable variation between individuals of a species, but if this variation is continuous, division into separate species would be based on subjective judgement. The interbreeding criterion cannot be relied upon either; two organisms which are dissimilar in some respects may be capable of breeding successfully, but this by itself does not mean they should be considered to be the same species.

Notwithstanding these difficulties, in practice we expect to find a considerable degree of resemblance between individuals of the same species; we also expect them to be capable of interbreeding; and we would hope for discontinuous variation in taxonomically-important characters between individuals of one species and individuals of another species.

The problem of defining the term 'species' is an expression of the nature of biological variation. Some groups of organisms are notoriously difficult to classify, and there may be a variety of opinions as to how the variation present should be categorized. For much of the time though, the complexities of taxonomy need not concern us unduly, but it is important to be aware of some of the issues involved.

Speciation

New species arise from within an existing species gene pool when individuals form a population that diverges sufficiently from the others so that the two groups can no longer interbreed. Fertile offspring can no longer be produced between members of the 'old' and the 'new' popula-tions. In other words, *gene flow* between the two populations ceases. Such *reproductive isolation* may occur in a number of ways. The most obvious way is through the separation of geographical ranges of two populations of the same species. When the distance between members of two populations is so great that they cannot possibly breed naturally with each other, they can diverge in response to different selective pressures and become increasingly dissimilar. However, if the two populations were in similar environments, and subject to similar selection pressures, they might not diverge very much. Nevertheless, geographical isolation is a means of curtailing gene flow and an important mechanism for speciation.

New species might arise from an old species, within the geographical range of the latter. Individuals could for example occupy different habitats or, in the case of animals, behavioural differences may prevent their mating. Reproductive isolation may be associated with differences in life cycle characteristics, in the timing of reproductive events for example, that prevent individuals from interbreeding.

Among plants, a particularly important isolating mechanism involves the spontaneous doubling of chromosomes. Sometimes, two closely related yet, distinct forms, *hybridize*, but the chromosomes from the two parents do not pair. However, a spontaneous doubling of chromosomes may occur producing two homologues, i.e. a pair of each type of chromosome present. The individuals which result are appropriately termed *polyploids*.

Species names

Each kind of organism known to science has a proper, or scientific, name. The name comprises two parts, both of which are Latin, or Latinized, and each of which should be written in italics or else underlined. The scientific name for the tree commonly known as the coast redwood in California is *Sequoia sempervirens*; the proper name for the orang-utan of Sumatra and Borneo is *Pongo pygmaeus*. The first word, the *generic* name, tells us to which *genus* (plural *genera*) the organism belongs. The second part, the *specific epithet*, completes the species name. (Note that the generic name begins with a capital letter but the specific epithet does not.) As specific epithets are used only in association with a generic name, the same specific epithet may be used in different genera. The names chosen for a species are often descriptive of the organism, named after a geographical locality, or after the discoverer. The number of species within genera varies widely; there may be only one, more usually there is a handful, but in exceptional cases over a thousand.

It was the Swedish natural historian Carolus Linnaeus (1707–78) who standardized the use of the *binomial system* for naming organisms. Previously, scientific names had typically comprised several parts; these were descriptive of the organisms, but the system was too cumbersome for general use. The binomial system has since been accepted throughout the scientific world. When written, a species name will often be followed by a capital letter, or a few letters, informing us of the *author*. The author is the person who first described that species. Many species names are followed by L. which stands for Linnaeus himself. It is useful to cite the author, particularly when dealing with 'difficult' taxonomic groups, because it removes doubt as to which kind of organism reference is being made.

Sometimes it is possible to recognize *subspecies* or *varieties* of a species. This could happen where two populations have been isolated geographically from each other and have evolved some slight differences in form, but their divergence is insufficient to warrant separate species status. If the study of a third population showed it to be intermediate in character between the other two, such that variation between the three populations is continuous, then the two subspecies categories would no longer be justified.

Common names for organisms are frequently employed informally. However, there is no universally recognized scheme for their use, such names tend to be used locally and inconsistently, and one name frequently covers more than one species. In the United States, 'robin' is usually taken to mean the species *Turdus migratorius*, but in Britain 'robin' generally means the species *Erithacus rubecula*. Common names are sometimes used for convenience, as in this book, but only where the point being made does not demand taxonomic precision.

How many species are there?

Notwithstanding the problem of defining the term 'species', the question 'how many species are there?' is an important one. Really there are two questions here: first, the number of species known to science and second, the number that actually exists on the planet. Perhaps surprisingly, there is no single inventory for named species. The distinguished American biologist Edward O Wilson has recently suggested that the number so far described is around 1.4 million. Estimates of the *total* number of species vary considerably, but all exceed by far the number of those so far described and named. A glance at the literature for the 1960s suggests a figure of around four million was then thought to be reasonable. However, more recent estimates indicate that upwards of twenty million, and perhaps as many as forty million, would be more realistic.

This hugely increased estimate over a relatively short time-span is due mainly to the recognition of the fantastic diversity of insects in the forests of the humid tropics. Sadly, the pressure on these environments from human activities is such that we may never know whether or not these recent estimates are correct. It is sobering to reflect that only a small proportion of the species with which we share the planet have been scientifically named and described.

The taxonomic hierarchy

Biological classification schemes are hierarchical, meaning that they comprise various levels, or taxonomic ranks, as shown here:

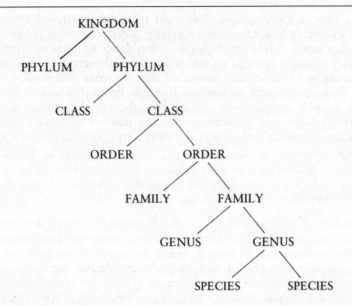

Thus, genera are grouped into families, families into orders and so on. There are in fact other taxonomic ranks in between those shown, but this scheme will be quite adequate for our purposes. As we ascend the hierarchy, the number of groups becomes smaller and smaller and there is less and less overall resemblance between the species comprising a group. Any particular taxonomic group, of whatever rank, can be referred to as a *taxon* (plural *taxa*).

The organisms comprising a single phylum all share certain characteristics, even though superficially they may seem very diverse. Such basic characteristics are very useful for determining long-term evolutionary lineages. For organisms other than animals, the term, division, has traditionally been used rather than phylum, but the latter is becoming much more widely used.

Taxonomic revisions

As pointed out at the beginning of this chapter, classification schemes reflect human interpretation of biological variation, so not surprisingly there is no single, universally agreed taxonomic scheme for all organisms. In fact, taxonomic revisions are continually being made as new information becomes available. Perhaps a study of previously neglected features argues for a new classification, perhaps new fossil evidence suggests a new phylogenetic scheme.

A taxonomic revision might involve the transfer of a species from one genus to another (necessitating a change in its generic name at least); two or more species may be 'amalgamated' or, conversely, one species may be split into two or more species. The coast redwood mentioned above provides an example. This taxon was formerly called *Taxodium semper-virens*; in other words, taxonomists felt its resemblance to other members of the genus *Taxodium* justified its inclusion within this genus. However closer examination of these taxa suggested that the coast redwood was sufficiently distinct to justify its removal from the genus *Taxodium*. It is now called *Sequoia sempervirens*. (Note that the specific epithet was retained.) For a large and 'difficult' genus, the number of species may vary depending on which authority is consulted; arguing strongly for use of the author's name.

Kingdoms of organisms

The highest formal taxonomic rank is the kingdom, of which five are now generally recognized. There is no doubt concerning the kingdom to which most organisms should belong, but a few types of organism are assigned to different kingdoms in different schemes. Again, the reason is simply that different characters are chosen as criteria for classification.

Faced with the problem of categorizing organisms, most people would immediately make a distinction between plants and animals, and have no difficulty in assigning most of the organisms around them to one or other of the two groups. The problem with a two-kingdom scheme, though, is that many organisms do not fit comfortably into either when we try to define the terms 'plant' and 'animal'. Most of these organisms are not visible to the naked eye, so until the first microscopes were used in the seventeenth century the problem did not exist. Even after the discovery of microorganisms, most seemed either more 'plant like' or more 'animal like' and received attention accordingly from either botanists or zoologists. One exceedingly important group of microorganisms, the bacteria, never fitted comfortably into either of the two major kingdoms and bacteriology developed to a large extent as a separate discipline, associated particularly with medicine.

As the biology of microorganisms became better understood it was clear that a two-kingdom scheme was unsatisfactory and so a third kingdom was proposed by some biologists. However, during the last few decades, largely because of the introduction of the electron microscope, microorganisms have been shown to vary in such fundamental ways that a further increase in the number of kingdoms was deemed to be necessary.

In fact, as mentioned frequently in previous chapters, two quite fundamentally different types of organism exist, the prokaryotes and the eukaryotes. Prokaryotic organisms comprise a separate kingdom, the Monera. Remember that all prokaryotic organisms are essentially unicellular whereas eukaryotic organisms may be unicellular or multicellular. Prokaryotic cells are very much simpler in terms of cellular organization, and they are also smaller than eukaryotic cells. Recall that the DNA of a prokaryotic cell forms a single, circular chromosome, whereas the

DNA in a eukaryotic cell is packaged, with associated protein, on several paired chromosomes within a membrane-bound nucleus. Prokaryotic cells do not have the various organelles found in most eukaryotes, although they carry out broadly similar functions. It may seem surprising, but in many fundamental ways the differences between a prokaryote and a eukaryotic unicellular organism, such as an amoeba, are far greater than the differences between an amoeba, and any multicellular organism. The bacteria, including the cyanobacteria, that make up the kingdom Monera play an exceedingly important role in the function of the biosphere and collectively they exhibit a greater range of metabolic diversity and versatility than all other types of organism combined.

All other unicellular organisms, together with all multicellular organisms, are eukaryotic. The kingdom Protista (or Protoctista) comprises all unicellular eukaryotes and a few primitive multicellular organisms that do not fit comfortably into any of the three wholly multicellular kingdoms. The term 'protist' may be unfamiliar, but we will now use it quite frequently. Important members of the Protista include the unicellular algae, which are the principal photosynthetic organisms in open water, multicellular algae (the 'seaweeds'), and the protozoa, a large group of unicellular organisms of which many, such as the malaria parasite, cause disease.

The remaining three kingdoms are almost exclusively multicellular. The fungi, despite their historically-close botanical associations, are sufficiently distinctive to justify a kingdom of their own. The plant kingdom is made up of multicellular, mostly 'green', plants including mosses and liverworts, clubmosses, horsetails, the ferns and the seed plants. The animal kingdom has a greater degree of uniformity than in the former two-kingdom scheme as it no longer contains any unicellular organisms and has also lost some of the more primitive multicellular organisms to the Protista.

Viruses are not organisms, and hence not assigned to any of the five kingdoms. They are, however, entities of considerable biological importance. A virus consists of a strand of nucleic acid and, usually, a protein coat. Replication of a virus can seemingly occur only within a living cell; the host cell's metabolic machinery being used to carry out the instructions coded for by the viral nucleic acid. Viruses can induce disease in the host organism; in fact, many diseases of crops and livestock, as well as of humans, including AIDS (acquired immune deficiency syndrome), are virally induced.

Further reading

Dawkins, R. (1988) *The Blind Watchmaker*. (1989) *The Selfish Gene*, 2nd edn. London, Penguin Books.
(Two finely crafted books on evolution; accessible and almost guaranteed to enthuse and enlighten.)
Dowdeswell, W.H. (1984) *Evolution – A modern synthesis*. London, Heinemann.
(Concise introduction to evolutionary processes.)
Margulis, L. and Schwartz, K.V. (1988) *Five Kingdoms – An illustrated guide to the phyla of life on Earth*, 2nd edn. New York, W.H. Freeman.
(Excellent reference to the diversity of life; strong on functional aspects.)
O'Neill, P. (1985) *Environmental Chemistry*. London, George Allen and Unwin.
(Very useful treatment of the subject with sympathetic introduction to some elementary chemistry.)
Postgate, J. (1986) *Microbes and Man*, 2nd edn. London, Penguin Books.
(Highly readable introduction to microbial life by a distinguished microbiologist.)
Rose, S. (1979) *The Chemistry of Life*, 2nd edn. London, Penguin Books.
(Rigorous and concise introduction to biochemistry.)

Organisms and ecosystems

Part 2 *Organisms and ecosystems*

During the first part of this book we paid particular attention to cellular processes because it is within cells that the really vital metabolic processes occur. Now we shift the level of focus and consider whole organisms and ecological systems, although we shall still be concerned with functional aspects.

The aim of this section is to provide a framework for the study of ecosystem function, regardless of type of ecosystem and regardless of spatial scale. In Chapters 7 and 8 we look more closely at the characteristics of autotrophic and heterotrophic organisms (a distinction we drew in Chapter 5) and consider what is meant by, and what factors influence, biological productivity. The use of the food chain as an organizational framework for ecosystem analysis is introduced in Chapter 9, together with some related topics. Finally, in Chapter 10, discussion centres on the transfer of chemical elements through the biosphere.

7 Autotrophic organisms and primary production

In Chapter 5 we categorized organisms as either autotrophic or hetero-trophic. Recall that autotrophic organisms manufacture energy-rich organic molecules from carbon dioxide using either sunlight (photosynthesis) or simple inorganic chemicals (chemosynthesis) as an energy source. Heterotrophic organisms in contrast require preformed organic molecules. So essentially organisms are either independent, or are dependent upon other organisms for their energy needs.

Photosynthesis is such a fundamentally important process in the biosphere that it merits further discussion. All the organic matter on our planet is derived ultimately from the photosynthetic process. In addition, all the free oxygen in the atmosphere and in water is produced from photosynthesis. Photosynthesis underpins all biological production, supplies us with our food needs, our fibre and most of our fuel. In this context it is worth noting that the substances we refer to as 'fossil fuels' are the much-altered products of photosynthesis from a few hundred million years ago.

The green plants, certain kinds of unicellular eukaryote, and some bacteria (including the cyanobacteria and those types that use hydrogen sulphide) are photosynthetic. In this chapter we look more closely at the process of photosynthesis, the factors that influence the rate of photosynthesis and the rate at which organic matter is produced. We then briefly return to that other mode of autotrophic nutrition known as chemosynthesis.

Photosynthesis and solar radiation

Solar energy is transmitted in the form of *electromagnetic waves* within a particular range of wavelengths. The wavelength range received by the Earth occupies just a small part of the whole *electromagnetic spectrum*, which extends from highly energetic gamma rays and X-rays to low energy radio waves (Fig. 7.1). The energy of electromagnetic waves varies according to the wavelength; the shorter the wavelength the greater the energy.

Electromagnetic radiation of only certain wavelengths is used in photosynthesis. This wavelength range, known as *photosynthetically*

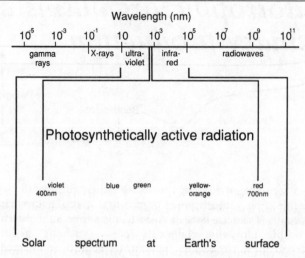

Fig. 7.1 The electromagnetic spectrum: the wavelength range of photosynthetically active radiation, which approximately coincides with light, occupies a very small part of the complete spectrum.

active radiation (PAR) extends approximately between 400–700 nanometres (nm), which is roughly equivalent to visible radiation or 'light'. Thus the expression 'light energy'. The energy within PAR is approximately half the total received at the Earth's surface. The energy received at the Earth's surface is only about half that at the outer atmosphere because of absorption by gases, scattering and reflection.

Much of the Earth's surface is covered by water of course. As the sun's electromagnetic waves pass through water, they are absorbed differentially. The 'visible' red, together with the infrared, wavebands are absorbed by the water itself in the first half metre or so, reducing the energy by about one half. In clear water the blue and green wavebands penetrate further than the red and yellow wavebands, but substances in the water affect considerably the rate at which the various wavelengths are attenuated.

As light is required for photosynthesis, the rate of light attenuation in water determines the depth to which photosynthesis can occur. That part of the water column in which there is sufficient light for photosynthesis is known as the *euphotic* or *photic* zone (Fig. 7.2). The lower limit of the euphotic zone varies considerably: it varies at any single location as sunlight changes and the turbidity of the water alters, and it varies spatially. In clear oceanic waters of the tropics photosynthesis may occur to depths in excess of 100 metres, but in most waters the euphotic zone is much less deep. In estuarine waters, which typically contain a considerable amount of material in suspension and in solution, the euphotic zone may not extend beyond a few metres. In relatively shallow water, of course, photosynthesis may occur on the bottom, i.e. in the *benthic* zone (Fig. 7.2).

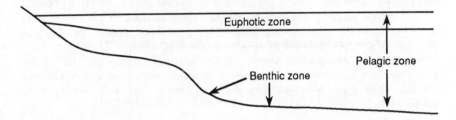

Fig. 7.2 Zonation in a body of water. The euphotic zone extends to that depth at which light energy is adequate for photosynthesis.

Although our primary concern here is with photosynthetically active radiation, some wavebands outside this range also have important biological effects. Parts of the infrared region trigger a number of biochemical reactions essential for function and development in some types of organism. Wavelengths in the ultraviolet part of the spectrum, particularly the shorter wavelengths, are biologically active. Ultraviolet wavelengths of less than 320 nm are absorbed by DNA and can induce mutations. The Earth's surface is screened from much of the incoming ultraviolet radiation, particularly these shorter wavelengths, by the ozone (O_3) in the lower stratosphere, so evidence for a decline in ozone concentration is a matter of considerable concern. Incidentally, atmospheric ozone is derived from oxygen, so its presence is due to photosynthesis.

The photosynthetic apparatus

In the photosynthetic process, radiant energy of particular wavelengths is absorbed by molecules of chlorophyll, the substance that provides photosynthetic tissue with its characteristic green colour. Although the singular is used here, there is more than one type of chlorophyll. Most plants contain two types of chlorophyll (known as chlorophyll$_a$ and chlorophyll$_b$) both of which differ from the chlorophyll molecules of photosynthetic bacteria. The fact that some plant leaves are tinged red or blue or yellow does not necessarily mean they contain no chlorophyll, merely that the chlorophyll is being masked by other pigments.

In the photosynthetic cells of eukaryotic organisms, the chlorophyll is located within organelles called *chloroplasts*. In the photosynthetic bacteria the chlorophyll is associated with the outer membrane of the cell (remember that prokaryotes have no cellular organelles). Absorption of radiant energy by chlorophyll in plants is strongest in the red and blue ends of the solar spectrum and weakest in the green and yellow bands

(Fig. 7.1), although radiant energy in these intermediate wavelengths is trapped by other, *accessory* pigments and passed on to the chlorophylls.

In the case of most plants, the carbon dioxide used in photosynthesis diffuses from the atmosphere into the intercellular spaces of the leaf through tiny pores, known as *stoma*, in the leaf surface. These pores are surrounded by *guard cells* whose water status determines the size of the pore; the whole apparatus is termed a *stomate* (plural *stomata*). Carbon dioxide then goes into solution before passing into cells where it is assimilated.

Gross and net photosynthesis

As discussed in Chapter 5, all work carried out within a cell involves an expenditure of energy. So while green plants are photosynthesizing they are also respiring, i.e. using the chemical products of photosynthesis in respiration. *Gross* photosynthesis refers to the total amount of energy captured, or carbon assimilated, before an allowance is made for respiration. The rate of photosynthesis after respiratory losses are taken into account is the *net* photosynthetic rate. Now if a plant uses up more energy in respiration than it gains in photosynthesis during a particular time interval then it will contain less energy, effectively weigh less, at the end of the period than at the beginning. Plants must therefore lose some weight, however small, during the night.

Not surprisingly, the amount of photosynthetic tissue (usually the total area of leaves) relative to the amount of living, but non-photosynthetic tissue, has an important influence on the rate at which a plant grows. A plant from which most of the leaves have been removed may not assimilate carbon fast enough to balance the inevitable respiratory losses from its non-photosynthetic parts such as roots and stems. The carbon economy is clearly crucial to, and more or less defines, a plant's performance.

Environmental effects on photosynthesis

Photosynthesis and light

The rate at which a leaf, or plant, assimilates carbon in photosynthesis varies according to the prevailing environmental conditions. We shall discuss first how photosynthesis varies in response to light. In the laboratory, and in the field, the *apparent* rate of photosynthesis can be determined by measuring the loss of carbon dioxide from an air stream passing over an enclosed leaf or plant.

Two *light response curves* for photosynthesis are depicted in Fig. 7.3(a). They show how the rate of photosynthesis changes as irradiance increases. 'Full sunlight' is equivalent to the maximum amount of sunlight likely to be encountered. First, consider the curve C_3. Notice

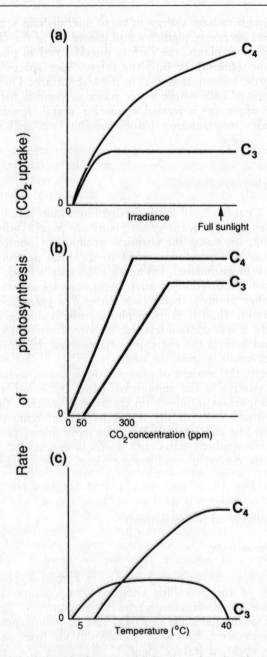

Fig. 7.3 General relationships between (a) light, (b) carbon dioxide, (c) temperature and the apparent rate of photosynthesis of plant leaves. The curves are not derived from actual data but show the characteristic differences between C_3 and C_4 plants. No values for carbon dioxide uptake are shown, but the scale is an arithmetic one.

that as the light level increases above the minimum necessary for photosynthesis, the rate of photosynthesis increases more or less linearly, suggesting that the availability of light energy is limiting the photosynthetic process. As irradiance increases further, the proportional increase in photosynthesis becomes less and less, until finally photosynthesis fails to respond at all to any further increase in light. The light level at which photosynthesis is at its maximum is known as the *light saturation point*: in this example the leaf 'light saturates' at around twenty-five per cent of full sunlight.

Curve C_4 in Fig. 7.3a represents a light response curve of a leaf from a plant with rather different photosynthetic characteristics. This leaf shows a broadly similar response to Leaf C_3 at low levels of irradiance but differs markedly at higher light levels. It does not 'light saturate'; rather, the rate of photosynthesis continues to increase up to full sunlight, although the curve is parabaloid in form rather than linear. Notice also that the photosynthetic rate at all light levels, except for the very lowest, is greater in Leaf C_4 than in Leaf C_3 and that this difference is accentuated as light levels increase beyond the point of light saturation for Leaf C_3.

The light response curve of most plants tends, *in general*, to be somewhat similar to one of the two curves in Fig. 7.3a. The labels C_3 and C_4 have a meaning; they refer to two distinct types of photosynthesis, and they are used also to refer to the plants themselves. Most plants can be categorized as either C_3 or C_4 plants depending on their photosynthetic characteristics.

The quantitative difference in response to light shown by the two types of plant is actually only one manifestation of the two types of photosynthesis. The title C_3 is used because the first stable product of this type of photosynthesis, a compound called phosphoglyceric acid, has three carbon atoms. The biochemical pathway by which carbon dioxide is reduced and phosphoglyceric acid is produced is known as the Calvin pathway, or Calvin cycle, after the scientist responsible for working out its details in the late 1940s and early 1950s. For this very important contribution, Calvin, who was working at the University of California at Berkeley, was awarded a Nobel prize.

Curve C_4 in Fig. 7.3a typifies a C_4 plant, so called because the first stable products of photosynthesis are compounds containing four carbon atoms (either malate or aspartate depending on species). The fact that some plants possess a different type of photosynthesis from that of C_3 plants was not appreciated until the 1960s. The C_4 biochemical pathway is sometimes referred to as the Hatch and Slack pathway after the two scientists who were largely responsible for its elucidation.

Photosynthesis in both C_3 and C_4 plants involves the C_3 (Calvin) pathway; in the leaves of C_4 plants, however, carbon is first 'fixed' as malate or aspartate, in outer *mesophyll* cells and these molecules are then transported to inner, *bundle sheath* cells where the carbon dioxide is released and biochemically 'fixed' again in the Calvin cycle. An outline scheme showing the main features of contrast between the two leaf types

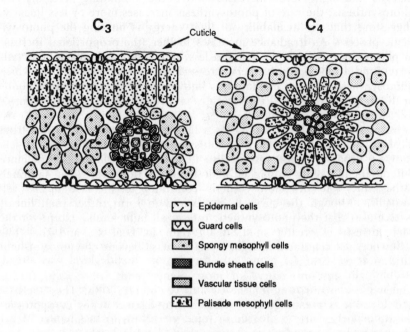

Fig. 7.4   Schematic layout of cells in C_3 and C_4 leaves in cross section. Dark patches within cells denote chloroplasts. Vascular tissue conveys nutrients, assimilates and water. The guard cells of the stomata surround pores through which gas exchange occurs. Although frequently represented as two-dimensional, cells should always be thought of as three-dimensional structures.

is shown in Fig. 7.4. The net effect of this anatomical and physiological specialization among cells is to produce a very high concentration of carbon dioxide in the bundle sheath cells.

There is an interesting geographical dimension to the distribution of C_3 and C_4 plants. C_4 plants, which are found principally within the tropics, subtropics and warm temperate zones, are usually associated with climates characterized by prolonged hot, dry periods, although even under these conditions C_3 plants are usually predominant. In cooler and/or more humid zones, virtually all the plant species are of the C_3 type. Among important food plants, wheat (*Triticum aestivum*), barley (*Hordeum vulgare*), rice (*Oryza sativa*), soya bean (*Glycine max*) and the potato (*Solanum tuberosum*) are C_3 species, whereas sugar cane (*Saccharum officinarum*) and corn or maize (*Zea mays*) are C_4 plants.

Photosynthesis and carbon dioxide

The biochemical and anatomical differences between C_3 and C_4 plants are associated with a difference in response to ambient (i.e. 'surrounding') carbon dioxide levels (Fig. 7.3b). Again, the curves represent hypothetical plants and there is variation within each group. Notice that for both C_3 and C_4 plants the relationship between carbon dioxide concentration and photosynthesis is initially linear, although the photosynthetic rate is greater for C_4 than for C_3 plants. There are two other important differences. First, C_4 leaves continue to take up carbon dioxide at concentrations down to practically zero. In contrast, C_3 leaves do not assimilate carbon dioxide below an ambient level of about fifty parts per million (ppm). Second, C_4 leaves are unable to respond to carbon dioxide in excess of about 300 ppm (which is somewhat less than the current atmospheric level), whereas C_3 leaves respond to carbon dioxide levels considerably in excess of this value. Thus commercial growers can boost production by generating carbon dioxide artificially in glasshouses. (Obviously their efforts would be in vain if they were growing C_4 plants and the carbon dioxide level was already around 300 ppm.)

The photosynthetic response of plants to carbon dioxide is currently of considerable relevance because of the upward trend in the concentration of atmospheric carbon dioxide (a topic we return to in Chapter 10). It is sometimes asserted that rising carbon dioxide levels will lead to a higher overall rate of photosynthesis. However, increasing carbon dioxide levels may well lead to some global warming and changes in rainfall patterns, both of which carry implications for photosynthesis. So it is not yet possible to make confident predictions about the effects of rising carbon dioxide levels on the biosphere.

A satisfactory explanation as to why C_3 and C_4 plants respond differently to ambient carbon dioxide concentrations is really beyond the scope of this book. However, it is useful to appreciate that one manifestation of this difference is that whereas C_3 leaves release carbon dioxide in the light, C_4 leaves do not. This is because of a biochemical pathway in C_3 leaves which breaks down some of the intermediate products of photosynthesis. This process, which is induced by light, is called *photorespiration*: it does not appear to serve any useful function and is sometimes described as 'wasteful'.

The reason that the carbon dioxide uptake by a C_3 leaf ceases when the ambient concentration falls below around 50 ppm is because the amount of carbon fixed in photosynthesis is insufficient to compensate for the carbon lost as a result of photorespiration. Now C_4 leaves do not release carbon dioxide in daylight, essentially because enzymes in the outer (mesophyll) cells are very efficient at incorporating any carbon dioxide present into malate or aspartate. Thus C_4 leaves continue fixing carbon at very low ambient carbon dioxide concentrations.

Photosynthesis and temperature

Two temperature response curves for photosynthesis are shown in Fig. 7.3c, and again we are dealing with hypothetical, but representative, cases. Typically, C_3 plants, particularly those from outside the tropics, have a comparatively low minimum threshold temperature for photosynthesis, and their photosynthetic rates are near their maximum over quite a wide temperature range. C_4 plants tend to have a higher minimum temperature for photosynthesis, but the rate of carbon fixation increases rapidly as the temperature rises. In addition, C_4 plants usually show higher optimum temperatures for photosynthesis, and at the optimum temperature they can assimilate carbon at a much greater rate than C_3 plants.

The effect of temperature on net photosynthesis is of course complicated by the fact that respiration is also temperature dependent. As the temperature increases, the rate of respiration rises. At very high temperatures, therefore, the rate at which carbon is used in respiration could be equal to the gross photosynthetic rate, in which case no growth would occur.

Photosynthesis and water

Photosynthesis is also responsive to moisture availability. It is the concentration of water (usually termed the *water potential*) in photosynthetic cells that is important. If the cellular water potential drops to a very low level, photosynthesis may cease altogether, but even before this point is reached the photosynthetic rate will fall. As we may expect, there are differences between plant species in this respect; some plants are much better equipped than others to cope with low cellular water potentials.

The capacity to withstand drought is not necessarily due to photosynthetic characteristics at low cellular water potentials, but may be due to features that permit plants to *avoid* low water potentials. Such features, which include deeply penetrating and extensive lateral roots, rolled leaves to minimize evaporative area, and water storage tissues, are commonly found in plants from dry environments.

The loss of water from a plant, principally in *transpiration*, is closely coupled with photosynthesis. This is because the principal routeway for water loss from a plant, the stomatal pores in the leaf surface, are used also by molecules of carbon dioxide diffusing into the leaf. As stomata open, the resistance to carbon dioxide and water vapour movement falls, i.e. *conductance* increases. Conversely, when stomata close, resistance to the movement of both gases is increased, i.e. conductance falls. As pointed out earlier, the size of these apertures is related to the water status of the surrounding guard cells, which is determined largely by environmental conditions. So, closed stomata may be good for water economy, but not necessarily conducive to rapid growth. On the other hand, wide-open stomata may permit a rapid rate of carbon dioxide diffusion and assimilation but at the expense of excessive water loss. The water and carbon economies of plants are therefore closely linked. Not surprisingly, plants from different environments respond in different ways to changes in water availability.

Another point worth making here is that as stomata behave like a variable resistance to water movement, the millions of stomata that exist in just a small area of vegetation may influence the rate at which water is lost from the soil, or from standing water, to the atmosphere.

Some plant species found in hot, dry environments possess a special type of photosynthesis, called *crassulacean acid metabolism* (CAM). Such plants take up carbon dioxide molecules during the night and chemically combine them (as molecules of aspartate); then, during the succeeding daylight period the carbon dioxide is released and used to manufacture carbohydrates. Such plants can keep their stomata closed during the hot daylight hours, thereby preventing excessive moisture loss. Although such plants use water very conservatively they have very low rates of carbon fixation.

The efficiency of photosynthesis

The efficiency with which solar energy is converted to chemical energy in photosynthesis is nowhere near 100 per cent. The maximum theoretical photosynthetic efficiency for an individual leaf under optimum conditions is about twenty-five per cent, but for communities of plants under field conditions much lower values are normal. Values of around ten per cent might be found for short periods, but averaged over a whole year photosynthetic efficiency is usually less than two per cent.

For the Earth's biota as a whole, photosynthetic efficiency is probably only a small fraction of one per cent. Part of the explanation for this low efficiency is that the cover of photosynthetic tissue over the Earth's surface is incomplete for a whole year, a fact that is evident to anyone who lives in a pronounced seasonal environment. Consequently, much incident radiation is 'wasted' as far as photosynthesis is concerned. Even if a green cover were maintained throughout the year, photosynthesis would be severely constrained by factors such as low temperatures or inadequate moisture outside the so-called 'growing season'. But even during the 'growing season' photosynthesis is limited by various environmental factors such as temperature, light, nutrients and water. In large measure, though, the very low efficiency with which solar energy is converted to chemical energy is explained by the photosynthetic process itself.

Photosynthetic *efficiency* is not the same as photosynthetic *rate*. In fact the efficiency with which radiant energy is converted to chemical energy declines as irradiance increases, even though the absolute photosynthetic rate continues to increase. It will be apparent from the light response curve C_3 in Fig. 7.3a, that beyond about twenty-five per cent of full sunlight, photosynthetic efficiency falls sharply, because photosynthesis does not respond to a further increase in irradiance.

Despite the low efficiency of solar energy conversion, the total amount of chemical energy made available annually to the biosphere by photosynthesis is enormous, exceeding by far the total energy consumption by the world's human population.

Primary production

Because photosynthetic (and chemosynthetic) organisms synthesize from carbon dioxide the organic building blocks on which all life depends they are referred to as *primary producers*. *Primary productivity* refers to the *rate* at which organic matter is produced. It is important to appreciate that the amount of plant material (plant *biomass*) in a particular area tells us very little about its rate of production. The weight of plant biomass in a hectare of forest could be several hundreds, or thousands, of tonnes, but unless we know the period of time it has taken for that amount of biomass to accumulate, how much has died and disappeared, and how much new leaf material is produced each year, we will know nothing about the primary productivity of the forest. Similarly, primary productivity in the oceans cannot be determined from the weight of the primary producers, principally tiny unicellular algae, whose biomass and longevity are only a tiny fraction of plants on land.

Primary productivity values are presented as units of weight (usually grams or tonnes of dry matter or carbon), or in energy terms (joules are now standard, but formerly calories were commonly used). These values must be expressed, for obvious reasons, on a unit area and unit time basis. Primary productivity measurements in energy terms permit the efficiency of solar energy capture by the community to be assessed, so long as a measure of solar energy input is available for the appropriate area and period. Productivity values are frequently presented on a calendar year basis, enabling comparison between community types and between geographical areas. In cases where interest centres on variation in productivity within a year, and how this variation correlates with particular environmental variables, data are presented for shorter intervals.

The distinction drawn earlier between gross and net photosynthesis applies also to primary production. *Gross* primary production is the total amount of carbon, or energy, fixed before an allowance is made for respiratory losses, while net primary production is what remains after respiration. So what we see, and what is actually available for use by other organisms, is net primary production.

A compilation of annual net primary productivity values are shown for comparative purposes in Table 7.1. In round figures, 4000 g m^{-2} year^{-1} seems to represent a convenient ceiling value, although values for most community types fall well below this amount. In general, primary productivity is lower in aquatic situations than on land. There is overlap though; productivity can be very high around coral reefs and in some estuaries, while on land, productivity tends to be low at high latitude,

Table 7.1. Suggested range of annual net primary production values for a variety of community types; based on published reports.

Community type	Productivity (g m^{-2} yr^{-1})
Tropical forest	1000–3500
Temperate forest	600–2500
Boreal forest	400–2000
Tropical grassland	500–4000
Temperate grassland	400–1500
Tundra and alpine	50– 400
Desert and semi-desert	10– 500
Agricultural land	100–5000
Swamps and marshes	800–3500
Lakes and streams	100–1500
Open ocean	50– 400
Nutrient-rich ocean	500–4000

high altitude and in dry desert areas. Notice also the wide range of values within each of the specified community types. Part of this variation may be attributed to real variation. In part though it may be due to methodology because there are enormous problems in measuring primary productivity accurately. It is particularly difficult to obtain reliable values for the productivity of underground plant parts.

Differences in primary production values between vegetation types, and between areas, are explicable mainly in terms of variation in environmental resources, notably light, water and nutrients, and in environmental conditions, particularly temperature. Also very important of course is how both resource availability and environmental conditions change over the year.

Some geographical trends in productivity are discernible. In humid climates, annual productivity of both lowland forest and grassland tends to increase from higher to lower latitudes, which correlates with the total amount of time during the year when photosynthesis occurs. Within the sub-humid, semi-arid and arid zones (which together account for over a third of the Earth's land area) there is a reasonable relationship between annual rainfall and primary productivity, indicating that moisture availability is the principal limiting factor.

In aquatic environments, nutrient availability is usually the critical factor, and of course it is nutrient availability in the euphotic zone that is important. In fresh-waters, phosphorus is generally the key limiting nutrient. In the open oceans it seems that nitrogen is usually the critical nutrient; there is not universal agreement on this point though and either phosphorus or iron may sometimes be the critical element. (We shall consider the factors that influence nutrient availability in water in Chapter 10.)

These are just generalizations; a real explanation for what is constraining productivity in a particular environment requires a close examination of that environment and the behaviour of the organisms within it. When discussing limiting factors, it is useful to distinguish between environmental factors that limit productivity on an annual basis and those that limit productivity at any instant in time. In dry deserts, moisture availability is the key limiting factor for annual productivity, but just after a rain storm water may not be the limiting factor; it could be air temperature or the availability of some nutrient in the soil. Similarly, on an annual basis, certain key nutrients limit productivity in the oceans surrounding Antarctica, but at any time during the long Antarctic winter, it is light that is the critical limiting factor.

Understanding the factors that affect photosynthesis and primary productivity is extremely important for human welfare. The science of agronomy is concerned largely with identifying the factors which constrain crop plant productivity and devising ways of overcoming them, either through environmental modification, e.g. drainage, irrigation, fertilization, or through the selection and breeding of plant genotypes that perform well under a particular set of environmental conditions.

Chemosynthesis

Chemosynthetic organisms can also be considered primary producers, even though the amount of organic material they manufacture is negligible. The chemosynthetic bacteria meet their energy needs by oxidizing inorganic chemicals scavenged from their environment and using this energy to manufacture organic compounds from carbon dioxide. In fact the biochemical pathways of carbon reduction are basically the same as in photosynthesis.

The chemosynthetic bacteria are classifed on the basis of the inorganic chemicals that they oxidize for their energy needs. Collectively they utilize a variety of such chemicals including ammonium ions (NH_4^+), nitrite ions (NO_2^-), ferrous iron (Fe^{3+}) and even hydrogen gas (H_2). When ammonium and nitrite are used as the oxidized substrates the processes can be summarized as follows:

$$2NH_4^+ + 3O_2 \rightarrow 2NO_2^- + 2H_2O + 4H^+ + energy$$

$$2NO_2^- + O_2 \rightarrow 2NO_3^- + energy$$

Chemosynthesis plays a vital role in the movement and transformation of certain chemical elements, notably nitrogen and sulphur, through the biosphere. For example, in soil and in water, ammonium ions (NH_4^+) are oxidized by certain types of chemosynthetic bacteria as represented above. The released nitrite ions (NO_2^-) are used by other sorts of bacteria which in turn produce nitrate ions (NO_3^-). Nitrate is the form in which most plants take up nitrogen from their environment. So these transformations are vital steps in the nitrogen cycle, a subject we return to in Chapter 10.

In most situations, chemosynthetic bacteria are unimportant for supplying new organic matter, but in the late 1970s such organisms were shown to play a key role as primary producers in the ecological communities associated with submarine hot springs (*hydrothermal vents*) on parts of the ocean floor. These features are confined to locations where materials emerge from beneath the Earth's surface to form new oceanic crust (see Fig. 13.5, page 156). The vents were discovered in an area known as the East Pacific Rise, but have since been found at sites on the Mid-Atlantic Ridge. Water, which has seeped into magma chambers below the crust's surface, is emitted from these vents at temperatures up to 300°C, and carries a variety of chemicals in solution. One of these chemicals is hydrogen sulphide. Bacteria found in and close to the vents oxidize the hydrogen sulphide to provide energy for the synthesis of organic molecules, using carbon dioxide dissolved in the sea water as a carbon source. A variety of organisms, including giant clams, tube worms and mussels depend on these bacteria for their energy needs. The bacteria have even been shown to reside within the cells of some of these organisms; consequently such animals do not need to eat the bacteria as they have their own intracellular community of primary producers. Such extraordinary mutualistic associations blur the distinction between autotrophs and heterotrophs.

8 Heterotrophic organisms and secondary production

Heterotrophic organisms cannot harness a nonbiological energy source to manufacture organic molecules from carbon dioxide. Instead they must consume preformed organic molecules which are later broken down, as discussed in Chapter 5, to meet their energy requirements. Heterotrophs may consume living tissue, eliminated organic substances, or dead remains, but all share a dependence on other organisms. In addition, only certain heterotrophs can produce amino acids; most therefore require a dietary supply of amino acids or proteins. All animals and fungi are heterotrophic, so too are the majority of protists (i.e. unicellular eukaryotes) and most groups of bacteria.

In this chapter we look a little closer at the heterotrophic way of life, paying particular attention to the fate of ingested food energy and what is meant by secondary production.

'Absorbers' and 'ingesters'

Most heterotrophic organisms can be assigned to one of two groups on the basis of the way they 'feed' and on their mode of digestion. One group, principally the heterotrophic bacteria and fungi, have no mouth parts; they release enzymes that catalyze the conversion of large organic molecules, such as cellulose, lignin, starch and proteins, to smaller entities such as sugars and amino acids. These molecules then pass into their cells where they are metabolized (Fig. 8.1(a). Organisms which acquire their food energy in this way can be thought of informally as 'absorbers'. They perform vital ecological roles, breaking down large organic molecules and making inorganic nutrients available again for uptake by other organisms.

In contrast to the absorbers, most members of the animal kingdom possess an internal space in which the enzymatic breakdown of large molecules occurs. Furthermore, most animals have some control over the materials that enter this internal space: they may be active and eat voluntarily, or, in the case of many sedentary aquatic animals, they may be passive feeders, filtering out food supplied by the surrounding water.

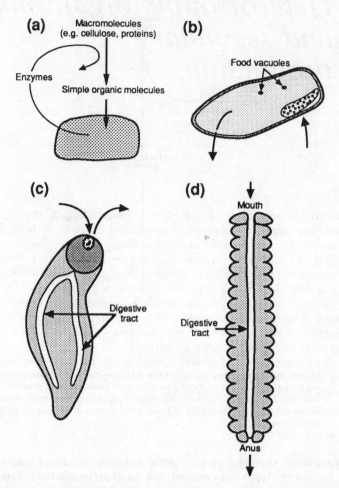

Fig. 8.1 Food procurement and digestion by heterotrophic organisms. (a) An 'absorber':
enzymes are released by bacterial and fungal cells into their local environment.
(b) A paramecium: this unicellular protist takes in and eliminates substances at
specific parts of the cell; unusually, digestion is intracellular. (c) A fluke, a type
of parasitic flatworm: the same orifice is used for ingestion and elimination.
(d) Earthworm: a tube like digestive tract with separate openings for ingestion
and elimination; the earthworm's digestive tract, which is actually differentiated
morphologically and functionally to a limited degree, provides a useful model
for most 'ingesters'.

Either way, such organisms might be termed 'ingesters'. Earthworms,
which have a tube-like internal space with a mouth at one end and an
anus at the other, serve as a good model for this type (Fig. 8.1d). In such
cases the enzymatic breakdown of food still occurs outside the cell; but
it is internal in the sense that the space in which it occurs is surrounded
by the body of the organism. This internal space is known variously as

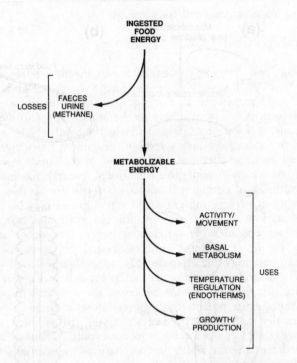

Fig. 8.2 Losses and uses of food energy ingested by an animal. Methane is generated by microbial metabolism in the digestive tract of some animals, notably ruminants and termites. Energy expended on thermal regulation (essentially mammals and birds) can be considered part of basal metabolism but of course varies according to temperature.

the *alimentary*, *digestive* or *gastrointestinal* tract or, more informally, the *gut*. In many types of animal the digestive tract is rather more complicated than that of an earthworm, but nevertheless this simple model will suffice for the majority of ingesting organisms. The flatworm shown in Fig. 8.1 uses the same orifice for ingesting food and for expelling undigested materials.

Some heterotrophic organisms do not fit easily into either the 'absorber' or 'ingester' categories. The unicellular paramecium (Fig. 8.1) absorbs food particles which then form food vacuoles that are gradually digested. Digestion occurs therefore within the cell, which is unusual, and undigested materials are released at a particular part of the cell surface.

Uses and losses of ingested food energy

Digestibility

The energy and nutrient content of food consumed by an animal may be lost or used in a variety of ways (Fig. 8.2). A variable proportion of ingested food passes out of the digestive tract unused, so its energy and nutrient content is not available to the animal. *Digestibility* refers to the proportion of food that passes across the wall of the digestive tract. Digestibility values vary considerably, depending particularly on the type of animal and the chemical composition of the food. For familiar farm animals, on which considerable nutritional research has been carried out, the digestibility of common foodstuffs is fairly predictable: for cattle, the digestibility of very young grass and cereal grains may be over eighty per cent but as low as forty per cent for old grass and cereal straw.

There are considerable differences between types of animal in their capacity to digest cellulose and associated polymers found in plants. Cellulose, you may recall (page 22), is a glucose-based polysaccharide which is the principal component of plant cell walls. For an animal to make use of a plant's cellular contents, the cell wall must be ruptured; and if the energy potential of the cellulose is to be used, it must be enzymatically converted (hydrolysis again) to simpler molecules. As plants age, their cellulose content increases; also, cellulose is often closely associated with lignin which further reduces the availability of cellulose and cellular contents.

To deal with the point about cellulose digestion first. It seems only a very, very few animal species produce the enzyme, *cellulase*, that breaks down cellulose. However, certain microorganisms, notably particular groups of bacteria, do produce this enzyme. Now some animals harbour microorganisms of this sort in their digestive tracts, and these break down cellulose in essentially the same way as their free-living counterparts. The products of this process are then potentially available to the host animal.

The association between cellulose-digesting microbes and their hosts is most highly developed for animals referred to as *ruminants*. Included within this group are numerous large herbivores such as deer, antelope and kangaroos, and many important domestic animals such as cattle, sheep, goats and camels. Ruminant animals have a multichambered and much enlarged stomach (Fig. 8.3) in which a community of cellulose-digesting bacteria and other microbes resides. Conditions in the rumen are anaerobic, and so it functions essentially as a fermentation chamber. Here, cellulose is broken down to sugars, and the sugars are converted to fatty acids which are absorbed across the gut wall. Ruminant animals also regurgitate partially digested food from the rumen to the mouth for further mastication, a phenomenon termed cud-chewing. Some herbivorous animals, including horses, zebras and rabbits, have no rumen, but they do have cellulose-digesting bacteria which reside towards the rear of the digestive tract.

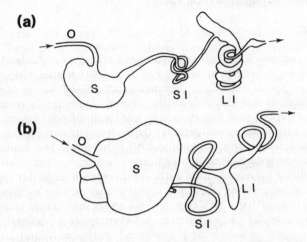

Fig. 8.3   Main features of digestive tract of (a) simple stomached mammal, and
(b) ruminant mammal. Symbols are; O oesophagus, S stomach, SI small
intestine, LI large intestine. The stomach of ruminants is a much enlarged,
multichambered organ occupied mainly by the rumen.

Associations between 'absorbing' microorganisms and their 'ingesting'
hosts do not necessarily involve large four-legged creatures. One very
important association is between termites (which are insects), and protozoans (unicellular protists). Termites devour wood and they are key
members of tropical and subtropical ecosystems. The digestion of woody
tissue is brought about not by termite enzymes though, but by enzymes
produced by the protozoan residents of the termite gut.

It should not be thought that herbivory is confined to animals with
cellulose-digesting microbes in their guts. Many types of herbivore
consume young plants, or plant parts, containing little cellulose so that
the cells are easily ruptured (much of the vegetable matter consumed by
humans is of this sort). The process of digestion begins in the mouth, but
some animals have another special compartment in which food is further
broken down and its digestibility increased. The gizzard serves this function in birds. Many herbivores, insects particularly, avoid the cellulose
problem by sucking the juices of plants or by burrowing inside cells.

The anaerobic processes that occur in the rumen lead to the production of some methane. Therefore, a small proportion of food energy
consumed by a ruminant animal will be lost in this form. A progressive
increase in atmospheric methane has been recorded over the last decade
or so, which may be at least partially attributable to an increase in world
cattle numbers. The significance of an increasing concentration of
atmospheric methane is that it is a very efficient 'greenhouse' gas, meaning that it traps outgoing longwave radiation (heat); its increase may
therefore contribute to global warming.

Digestible and metabolizable energy

Molecules that cross the wall of the digestive tract are carried to the parts of the body where they are used in metabolism. A small fraction of the digestible energy is lost as urine or other by-products of metabolism are released. The remainder is termed *metabolizable* energy. So for what purposes is metabolizable energy used by animals? The essential metabolic processes that keep an animal alive, e.g. pumping blood, transmitting nerve impulses and regulating cellular processes, are referred to collectively as *basal metabolism*. Basal metabolism is always an important item in an animal's energy budget, accounting for a significant proportion of metabolizable energy.

The minimum rate of energy expenditure for all essential metabolic processes is known as the *basal metabolic rate*. Although difficult to measure accurately, basal metabolic rate can be approximated by measuring the amount of oxygen consumed, the carbon dioxide released, or the heat generated by an organism at rest in an appropriate 'normal' temperature. For most organisms, basal metabolic rate changes with the stage of development: in humans it declines, on a unit weight basis, with age. In general, the smaller the animal the greater its metabolic rate *per unit of weight*. So, as a proportion of their body weight, smaller animals need to eat more food than larger animals, and in fact the amount of food consumed daily by some active small animals must exceed body weight. However the *total* amount of energy expended in basal metabolism increases more or less linearly with an increase in size.

Metabolic rate is affected by temperature, simply because the rate at which chemical processes proceed is temperature dependent. However some organisms, notably mammals and birds, use metabolic energy to maintain body temperature within a comparatively narrow range, and collectively reveal a variety of mechanisms for doing so. Because their body temperature is regulated internally, and kept fairly constant, such organisms are called *endotherms*, or *homeotherms*. The term 'warm-blooded' is used informally because body temperature is usually above that of their local environment. The more the external temperature departs from the 'normal' range, the greater the energy expended in thermal control, although of course there are limits to the temperature rises and falls that an endotherm can tolerate.

The vast majority of organisms do not use part of their metabolic energy to regulate their body temperature. Energy-demanding activities, and behaviour patterns, may serve to raise body temperatures a little but, in general, body temperature tends towards that of the environment. Such organisms are described as *ectothermic, poikilothermic*, or more loosely, 'cold-blooded'. At low temperatures, many ectotherms behave rather sluggishly; endotherms in contrast can remain active, which can give them a considerable advantage. However, it should not be thought that ectotherms are poorly adapted in general; the fact that food energy is not required for temperature regulation may be beneficial when food is in short supply. In fact body temperature of some endotherms drops at certain times, usually when food is scarce. Hibernating bears provide an example of such an adaptation.

All animals expend some energy on activity, the nature of the activity varying greatly of course between different sorts of animal. The greater the amount of activity the greater the energy expenditure, and therefore the greater the energy intake necessary to meet the demand. For mobile organisms, a high proportion of the energy budget may be expended in movement, much of it for food procurement.

Growth of animals

We have not yet considered growth of animals, by which we mean new production. Growth can occur only if the food energy ingested exceeds the energy lost and used in the various ways just described. Animal nutritionists employ the useful concept of *maintenance* to refer to a diet which keeps the weight of an animal constant; in other words, intake is equal to loss and expenditure during a particular time interval. Food energy in excess of the maintenance amount is available for growth, but it is not used for growth with 100 per cent efficiency. So, if the maintenance requirement is exceeded by 500 units of energy during a given time interval, we will not find 500 units of energy as new production during the same interval because there is an energetic cost associated with producing new tissue. The rate of growth depends on the efficiency with which food energy in excess of maintenance is used for synthesizing various chemical substances. It is not only food energy that is important here, but other aspects of the diet, particularly the quality and quantity of protein consumed.

Growth may be represented by a growing foetus, by milk or eggs as well as by muscle, fat, bone, hair or wool. Now if a pregnant animal, say a cow which is in calf, is fed less than the maintenance requirement, growth of the foetus will not automatically cease. Rather, the cow's own energy and nutrient reserves will be mobilized to meet the demands of the developing foetus. It is quite common for animals to lose body weight during times of the year when food supply is low, and during these periods the females may well be pregnant. Pregnant sheep, overwintering in the relatively harsh climate of the British uplands, provide an example of this situation.

The value of a diet in terms of new production can be determined under experimental conditions. Such work, which involves measurements of food intake, digestibility and growth, is routinely carried out on domesticated livestock in animal research institutes. There is much less information for undomesticated animals. However, a few years ago W.F. Humphreys of the University of Bath in England compiled and analyzed much of the data then available on this subject. Some of his findings are presented in Table 8.1. In this table the energy content of new growth (production) is expressed as a proportion of the total energy expended by the animal (respiration). The values therefore refer to metabolizable

Table 8.1   The efficiency with which assimilated food energy is used for productive purposes in several groups of animals. In each case values given are means from a number of studies. (Data compiled by W. Humphreys.)

Organism type	Efficiency (%)
(Homeotherms)	
Insectivores	0.9
Birds	1.3
Small mammals	1.5
Other mammals	3.1
(Poikilotherms)	
Fish and social insects	10.0
Invertebrates other than insects:	
Herbivores	20.8
Carnivores	27.6
Detritivores	36.2
Insects (non-social)	
Herbivores	38.8
Detritivores	47.0
Carnivores	55.6

energy, and not ingested energy, some of which is lost as faeces, urine or other excreted materials.

Clearly, there are big differences between types of animal. Notice that homeothermic animals have lower food conversion efficiencies than poikilotherms, presumably because of the energetic costs of regulating body temperature. Within the homeotherms, four distinct groups – insectivores, birds, small mammals and other mammals – emerged. Within the poikilotherms, fish and social insects comprise one group, the non-social insects comprise another group, and invertebrates other than insects yet a third group. The herbivorous invertebrates tend to have lower assimilation efficiencies than either carnivorous or detritivorous invertebrates, suggesting that plant material is converted into new animal tissue less efficiently than either animal tissue or dead organic matter.

Notice that new growth accounts for only a small, often very small, proportion of the food energy assimilated or ingested. It is in fact only in farm livestock that are being fed on very high energy and protein diets that production of large animals rises to perhaps twenty to thirty per cent of the ingested energy. Because basal metabolism always accounts for a large proportion of consumed food energy, high productivity from animals requires a high plane of nutrition. Modern farm livestock have of course been artificially selected (page 43) for their capacity to utilize a high energy and protein diet for producing milk, meat or eggs. In some situations, perhaps where future production rather than short-term yield is the aim, it could be advantageous to keep the maximum number of animals on a near-maintenance diet rather than a few animals on a high plane of nutrition. The course pursued will depend on the objectives. Moreover, animals have considerable cultural value in many societies and short-term economic gain may not be the major consideration.

Earlier, we made the point that protein quantity and quality were important dietary components for animals. Now although animals cannot themselves manufacture amino acids, some animals provide a suitable home for bacteria that can do so. This is another feature of the ruminant animals. Because of their bacterial residents, such animals are more or less independent of dietary protein, provided they have an alternative source of nitrogen in the diet. Ruminant animals, such as the camel, can survive on a diet consisting of poor plant material, not only because the rumen bacteria digest cellulose, but because they produce amino acids and hence protein. Such animals will not usually thrive on such a diet but they can survive, produce milk and blood, and reproduce.

Other heterotrophs

So far we have been discussing animal nutrition, but essentially the same principles apply to other heterotrophic organisms. In fact in terms of efficiency with which food energy is used for productive purposes there is no doubt that heterotrophic microorganisms are far superior to animals. Like animals, fungi and heterotrophic bacteria have a long history of use by human societies. Fungi may be consumed directly, as mushrooms are; microbes may be cultured because of the enzymes they produce, as in the case of yeasts which convert sugar to alcohol. Moreover, the high efficiency of food conversion by these organisms makes them potentially very useful for producing food for both humans and animals, particularly when waste materials are used as an energy substrate. Bacteria and fungi have particular potential because of their capacity to synthesize amino acids, and hence protein. Some animal-feed manufacturers use bacteria to produce microbial biomass which is then fed to farm animals. Moreover, human foods based on fungal biomass are being introduced to the market in some countries and could possibly become familiar dietary items in the future.

Secondary production

Secondary productivity is defined as the rate of organic matter accumulation by heterotrophic organisms, and as for primary productivity, its values are presented in dry weight or energy terms, per unit area per unit time. It is very difficult to obtain reliable estimates for secondary productivity under field conditions, particularly when an estimate of total secondary productivity is required. Part of the problem lies with the difficulty of determining population abundance for each species present, but many heterotrophic organisms are simply not amenable to such study in the field on account of their very small size or the habitats they occupy. Indeed for some types of organism, such as bacteria and fungi, the notion of number of individuals is rather meaningless and an alternative approach is necessary.

The task of measuring secondary productivity is simplified, although still subject to considerable problems, if data on a single animal species are required. In such a case, field studies of abundance and age and size distributions could be complemented by laboratory studies on individuals in an attempt to obtain a reasonable estimate of the true value. Measurements of secondary productivity provide useful insights into the way energy is transferred through ecological systems (a theme we explore further in the next chapter). Such data can inform decisions about the management of animal resources. A value for the annual production of a fish species could be used, for example, to set an upper limit to the permissible catch.

9 Food chains: the transfer of energy and matter through ecosystems

In Chapters 7 and 8 we discussed aspects of the autotrophic and heterotrophic ways of life and explained what is meant by primary and secondary production. Now we turn our attention to whole communities, and consider how chemical energy and nutrients are transferred as one organism obtains its nutritional needs from another, or from organic remains. As chemical energy is bound up within organic molecules the transfer of energy and chemical elements is closely coupled. When one organism consumes another, chemical energy is transferred together with a range of chemical elements; always carbon, hydrogen and oxygen of course, but mineral elements such as nitrogen, phosphorus, calcium and potassium as well. Such movements provide an important conceptual framework for the description, analysis and comparison of ecological systems and have been a major focus of ecological study for several decades.

Feeding relationships

Feeding relationships – roughly what consumes what and in what quantities – are central to the organization and function of ecological communities. The diet of an animal is a key feature of its ecology, and indeed heterotrophs are labelled according to their feeding habits. Thus herbivores use plants as a source of food, carnivores use other animals and omnivores consume both plant and animal material. Scavengers feed principally on animals that have recently died while detritivores consume organic remains. Expressions such as insectivore and frugivore (fruit eating) characterize diet even more specifically.

The term *predation* is a convenient one to use whenever energy and nutrients are transferred from one living organism to another. Predation does not necessarily bring about the death of the *prey* organism. When plants are grazed by large herbivores, for example, typically only a part of the leaf tissue of an individual plant will be consumed. Other predators of plants, such as birds and insects, may consume a proportion of the fruit or seeds, or in the case of insects, suck some sap from the

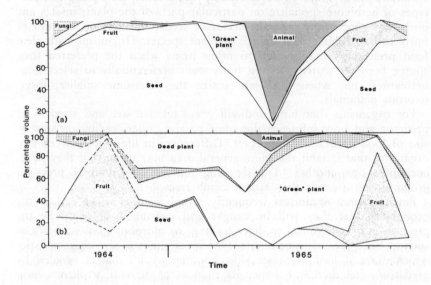

Fig. 9.1   Results from a study of seasonal dietary variation (a) wood mice and (b) bank
voles in an English wood. Both animals are primarily herbivorous but during
spring animal tissue is consumed by wood mice. The time intervals are months.

stem of a plant. Parasites (organisms that live on, and obtain their nutritional needs from, another living organism) do not necessarily kill their
hosts; indeed, death of the host is unlikely to be in the parasite's
interests. When parasitism is associated with the host's death it is quite
likely that the parasite contributed to death only in so far as it weakened
the host and increased its vulnerability to other stresses. In other cases,
and this applies particularly to insects which lay eggs in the juvenile
stages of other insects, parasitism results in almost certain death of the
host organism. The key point is that when energy and nutrients are
transferred from one living organism to another, the prey organism is
affected to a degree which depends upon the intensity and the nature of
the predation.

   As an animal grows and develops, its diet may change significantly.
The composition of an organism's diet may also change seasonally as the
availability of suitable food sources changes. Examples are shown in Fig.
9.1. The degree to which heterotrophs specialize with respect to their
food varies considerably between species; hence the terms 'generalist' and
'specialist' are useful ones for comparing dietary behaviour. Parasitic
organisms frequently display a high degree of host specificity; a parasitic

species may be found in just one location in one kind of host. Among herbivores there is considerable variation in dietary specialization. Domestic sheep graze more selectively than cattle for example. Many types of herbivore specialize on particular parts of the plant; insects and birds may concentrate on fruit, nectar or seeds, and may or may not be fairly catholic about their choice of plant species. Organisms with clear food preferences may switch to other items when the preferred food source becomes scarce. African lions seem preferentially to select large herbivores, but when these are scarce they consume smaller, insectivorous mammals.

For organisms that hunt and kill prey, relative size and strength of predator and prey influence diet. For passive, filter feeding organisms, size of food particles is important. Differences in life-style between two organisms that inhabit the same general area may mean that they never encounter one another. Tree-dwelling animals are unlikely prey for ground-living animals that cannot climb trees.

Potential prey organisms frequently reveal features which reduce the probability that they will be caught and consumed. It could be the production of a chemical to deter feeding, or morphological adaptations such as the spines of some plants. One species may evolve characteristics which make it look, or taste, like another species that is avoided by predators, and thereby enhance its chances of survival, a phenomenon known as *mimicry*.

Determining food type

A variety of approaches are used to determine the diets of heterotrophic organisms. The principal dietary components of some animals may be apparent from observation. We know African lions eat wildebeest and other large herbivores because we can watch them do so. It is clear that certain insects favour particular plant species as a source of food because we can watch them boring or sucking or grazing on these plants. Observation alone, however, will not permit the contributions of different dietary items to be quantified. In any case, some organisms are very difficult to observe because of their minute size or the habitats they occupy.

An animal's diet may be assessed by analysis of its gut contents. Problems with this approach are that it necessarily involves sacrifice of the animal and different food items are likely to be digested and removed from the gut at different rates, so care must be taken in interpreting the results from such investigations. Nevertheless, this method is widely used, particularly for organisms such as fish and other aquatic animals, whose feeding habits are difficult to observe but which are large enough to be amenable for such study.

Feeding habits may be investigated in a more experimental way. After 'labelling' a food source, perhaps a plant, with a small quantity of a radioisotope, organisms in the vicinity can be trapped and their radioactivity levels measured as a guide to what they are eating. Alternatively, food preferences may be assessed by providing an organism with a variety of foodstuffs and measuring its consumption of each. In discussing food preferences it is not only the 'ingesters' that need to be considered, but also bacteria and fungi. Species of such organisms can also show marked preferences for certain substrates, and these can be determined by observation and by experiment.

The effects of predators on prey

An important question concerning predation is the extent to which it affects abundance within prey populations. We have made the point that predation does not necessarily bring about the death of the prey organism, but clearly there are examples of predation which do result in prey death; it is this type of predation with which we are primarily concerned here. While discussing evolution in Chapter 4 we said that an intrinsic characteristic of all populations is a potential for growth in numbers. But in practice clearly only a fraction, often only a tiny fraction, of offspring survives to reproduce. So of the various factors that could contribute to the control of abundance, predation is at least a candidate.

Evidence for the efficacy of predation is provided by certain successful examples of biological control whereby the introduction of a 'natural enemy' has reduced the abundance of an unwanted species. However it would be unwise to conclude from such rather limited evidence, obtained from somewhat artificial situations, that predation is always a major factor in determining prey abundance. Indeed many attempts at biological control have been unsuccessful. Predation is a *potentially* important factor in controlling abundance; in some situations it may be very important, but of limited significance in others. To complicate matters, within particular predator-prey systems the influence of predation may vary from year to year.

In situations where death is not a direct consequence of predation, it may be that predation contributes to the control of prey abundance by affecting prey organisms in such a way as to increase their vulnerability to other factors, such as adverse weather and disease, and to reduce their capacity to compete with species with similar life-styles.

So generalizations concerning the contribution that predation makes to the control of prey abundance are best avoided: we shall note the very complex nature of this subject and the fact that it has long been a source of controversy among ecologists. The truth is probably that its significance varies considerably from situation to situation.

We can switch the question of course, and enquire how the availability of prey affects the abundance of predator populations. It is easy to demonstrate experimentally the considerable effect that food quantity and quality have on an organism's performance, including its reproductive performance, but very much more difficult to assess the influence of food supply over a long period under more natural conditions. So again, generalizations are not possible and likely to be misleading. Probably it is safe to say that the availability of food is of major significance in determining the abundance of some predators some of the time.

Decomposition

Decomposition means the breakdown of detritus. We introduce this important topic here because it involves the transfer of energy and nutrients between organisms and is central to any discussion of energy flow in ecosystems. Decomposition is essentially a biochemical process, under enzymatic control, but the physical processes of weathering and leaching also contribute. Although decomposition refers specifically to the loss of detritus, it should be appreciated that *all* living organisms contribute to the breakdown of organic matter, simply because all organisms, not just those associated with decomposition, transform chemical energy to heat.

The bulk of the detritus in most situations is made up of plant material; the remains of other organisms and the materials they release usually make a negligible contribution. In some coastal environments dead remains consist predominantly of 'shells' of marine organisms, but this is an exceptional case. The detritus generated within a defined area is known as the *autochthonous* component, while that imported is termed the *allochthonous* component. Allochthonous material frequently comprises the major part of the organic matter in ponds, stretches of running water and some estuaries.

The biological processing of detritus usually follows a fairly predictable pattern within a particular type of ecological community and it involves a variety of organisms. On land, leaf litter is usually rapidly colonized by fungi and also consumed by a variety of invertebrates such as insects (including ants), millipedes, snails and earthworms. Fungi are particularly important for the decomposition of woody material. Many types of fungus produce long, filamentous tissues called *hyphae* from which enzymes are released as they ramify through the dead tissues. 'Bracket' fungi, some of which grow to a considerable size, are often conspicuous on standing and fallen dead tree trunks. Heterotrophic types of bacteria are also major contributors to decomposition in most situations.

In aquatic environments, organic matter is usually much less concentrated than on land, and it occurs in a variety of chemical forms. A distinction is conventionally made between dissolved organic carbon, which accounts for most of the organic carbon, and particulate organic carbon, which covers a wide size-range. The organic carbon in water derives from dead organisms, from faecal pellets and various other eliminated substances, and much of it is processed first by bacteria.

Inspection of a soil profile will usually reveal a gradient in the appearance of organic matter, from the relatively unchanged plant litter on the surface through to the rather amorphous organic material which merges with the mineral layers. Organic matter associated with soil is commonly known as *humus*, and different types of humus characterize different soil types and different environments. Thus *mull* humus is relatively uniform, while *mor* humus is characterized by fairly distinct layers of organic matter in various stages of decomposition and transformation. Within the mineral horizons the organic matter is generally dispersed, although the identity of some organic remains may still be evident.

Decomposition also involves vitally important transformation of mineral elements from the organic form in which they are held to the inorganic form in which they can be utilized by plants and certain other organisms. We shall look more closely at this process, termed *mineralization*, in the next chapter.

The rate of decomposition

The amount of organic matter present at the soil surface, or on the bed of a water body, depends on the balance between the rate of addition of dead material and the rate at which it is decomposed. In situations where little organic matter is imported, the rate at which detritus is added should be related to primary productivity, although of course a proportion of the organic matter fixed by photosynthesis may accumulate as biomass, as during the growth of a stand of trees.

The amount of detritus present varies considerably between different community types and between different climates. For forests, amounts of detritus tend to decline from high latitude coniferous forest, through temperate forest to tropical forests. However, the amount of detritus should not be taken as an indication of the amount of organic matter added annually. On the warm, and frequently moist, tropical forest floor, biological activity is high and the large amount of material added each year is rapidly processed. In high latitude forests, in contrast, the relatively small amount of organic matter that drops to the forest floor each year is processed more slowly.

Rate of decomposition can be estimated by enclosing some detritus of known weight in small mesh bags, placing these in the field and periodically weighing them after their contents have been dried. By using mesh of different sizes, the contribution to decomposition of organisms of various sizes can be assessed. As the breakdown of detritus is principally a biological process, the rate of carbon dioxide generation may also be used as a measure of the rate of decomposition, although it should be remembered that organisms not involved in decomposition may simultaneously be generating carbon dioxide.

A number of factors combine and interact to determine the rate at which detritus is decomposed (Fig. 9.2). The type and relative abundance of the decomposer organisms present are important, and so too are the chemical and physical characteristics of the detritus. It is quite obvious that leaf litter decomposes much more rapidly than woody tissue, but the

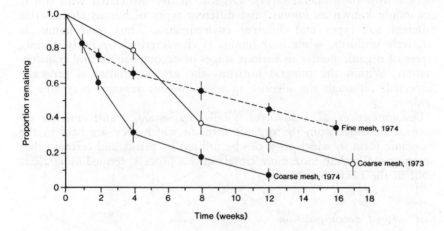

Fig. 9.2 Decomposition. The loss of litter with time from mesh bags placed on the ground in an area of shifting cultivation in Nigeria. Note effect of mesh size on the rate of loss and the difference between years.

rate of leaf decomposition varies between species due to differences in leaf chemistry. Environmental factors exert a major influence on the rate of decomposition because biological activity responds to changes in temperature, moisture, aeration and nutrient availability. In seasonal climates the rate of decomposition varies over the year, and at times may cease altogether, as for example during cold winters.

Decomposition tends to be stimulated by increased aeration. When soils are cultivated for the first time, a rapid loss of organic matter usually ensues before a new equilibrium between the rate of organic matter accumulation and its loss is established. Conversely, a reduction in aeration inhibits decomposition.

Organic matter accumulates on land, and at the bottom of a water column, when detritus additions exceed decomposition for a prolonged period. *Peat* comprises the accumulated, and physically transformed, remains of plant material. It forms in situations where prolonged waterlogging drastically lowers the oxygen concentration and inhibits decomposition, an effect that is often exacerbated by long, cold winters. Decomposition can still proceed under anaerobic conditions, as we should be aware from the discussion in Chapter 5, but at a considerably reduced rate.

In aquatic ecosystems, oxygen consumption is determined to a large extent by the level of activity of decomposer organisms, particularly bacteria. By stimulating bacterial activity, large loadings of dead organic matter can bring about a marked depletion of oxygen, leading to a deterioration in the ecological and amenity value of a water body. Lakes, rivers and estuaries that receive large quantities of sewage effluent often show marked oxygen depletions. A similar phenomenon occurs where

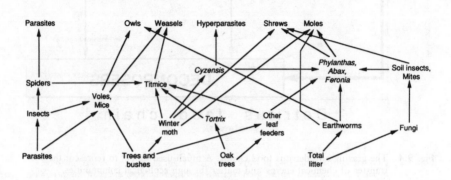

Fig. 9.3 A food web for a wood in southern England. Arrows represent the direction of flow of energy and nutrients between organisms.

nutrient loadings are high (in fact these nutrients may be released from sewage effluent), because primary productivity is closely related to the availability of nutrients, particularly phosphorus and nitrogen. Severe and prolonged oxygen depletion in water may lead to an accumulation of organic matter on the bottom, and anaerobic water bodies can be unpleasantly odorous.

Biological decay processes are harnessed in effluent treatment works for the breakdown of organic wastes. The aim is to provide an environment conducive to decomposition, but of course the efficiency of the process is affected by environmental conditions such as temperature. Also, biological activity may be inhibited by metabolic poisons that enter the effluent treatment works along with human sewage.

Food chains and food webs

The term *food chain* signifies the transfer of energy and matter between organisms. The food chain is a rather abstract concept, it really only exists as a diagram, as in Fig. 9.3, with the arrows showing the direction of movement of energy and materials between the indicated organisms.

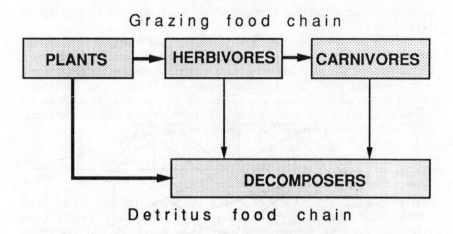

Fig. 9.4 The grazing and detritus food chains. A much-used model to represent the transfer of chemical energy and matter through ecological communities.

We talk of 'marine food chains' and 'arctic food chains' when referring to the typical patterns of movement of energy and materials between organisms in such environments, and the 'human food chain' when considering human dietary sources and the links involved prior to consumption by humans. The term *food web*, also widely used, reflects the numerous pathways of food transfer that frequently characterize ecological communities.

In drawing up a framework with which to examine energy and nutrient flow through a community, it is customary to distinguish a pathway based on living primary producers, the grazing food chain, from one based on detritus, as in Fig. 9.4. In this example, the grazing food chain comprises three 'stages', referred to as *trophic levels*. Examples of grazing food chains readily come to mind, perhaps the most vivid being those of African grasslands where plants are grazed by groups of large herbivores which are in turn the principal source of food for carnivorous animals.

The detritus food chain is usually far more difficult to study and analyze than the grazing food chain. Members of the detritus food chain are often extremely small and they tend to inhabit locations that are awkward to study, such as the soil or aquatic sediments. Furthermore, they interact in complex ways. Accordingly, the decomposers are frequently lumped together in a single box as in Fig. 9.4.

It is important to appreciate that most of the basic principles relating to food transfer through the grazing and the detritus food chains are essentially similar. However, one important difference between the grazing and detritus food chains is that whereas herbivore activity can influence the amount of energy and nutrients available to the grazing food chain, simply by affecting primary productivity, detritus feeders have no influence on the rate at which food material is supplied to them.

In nearly all situations the flow of energy and nutrients through the detritus food chain is much greater than through the grazing food chain. This is because the proportion of plant material consumed while it is still alive is usually much less than that which dies before being consumed. Furthermore, all organisms contribute to the detritus food chain on death, and may do so while living. Thus the arrows linking the herbivores and carnivores to the detritivores in Fig. 9.4.

Herbivore activity

Herbivores come in a variety of shapes and sizes, and they exploit their food in many different ways. Insects are major herbivores in virtually all terrestrial situations, sometimes in association with fruit- and seed-eating birds and small mammals. Large, hoofed mammals, including domesticated livestock, are often the most conspicuous herbivores on grasslands, but even in grassland communities the less obvious insects and soil-living invertebrates usually make a significant contribution to the amount of plant tissue consumed. In open water, tiny crustacean animals, which largely make up the *zooplankton*, are major primary consumers. In some situations, herbivorous fish and even aquatic mammals, may be the principal herbivores.

The proportion of annual primary production that is consumed by herbivores varies, but as just pointed out is usually less than the amount that dies before being used by heterotrophs. In natural and semi-natural grasslands, herbivores typically account for ten to thirty per cent of annual primary production, but in forests the value is probably usually less than ten per cent. In open water, where the principal primary producers are unicellular algae, values tend to exceed those typical for terrestrial environments. Whereas the whole of a unicellular alga in water is 'available' to a grazing organism, a significant portion of most land plants is situated underground and not easily exploited by grazing animals on the surface. Furthermore much new growth on land may be of fibrous material such as cellulose and lignin which is unpalatable to most herbivores.

The primary consumption values suggested above serve only as a guide; in practice values are likely to fluctuate, sometimes very markedly. In semi-arid areas of northern Africa, the abundance of locusts periodically increases dramatically, and local primary consumption values rise accordingly. Some species of forest insect, not very conspicuous in most years, periodically undergo a huge increase in abundance and cause serious defoliation of trees. One such insect is the commonly-named spruce budworm which has caused massive defoliation of coniferous trees in eastern Canada. Overgrazing by domestic livestock, which afflicts much of the world's dry zones, exemplifies chronically high primary consumption.

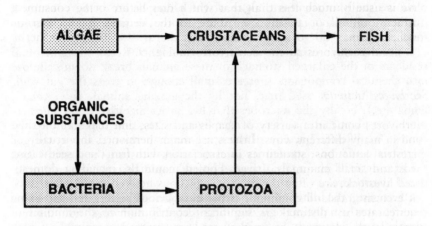

Fig. 9.5 Part of a marine food chain showing the problem of separating grazing and
detritus pathways. Crustaceans are consuming the primary producers but also
organisms in the detritus pathway which are partly sustained by organic
substances released by the algae. It is difficult in such a situation to calculate the
efficiency of energy transfer in the 'grazing food chain'.

The complex nature of food chains

The diagram in Fig. 9.4 provides a useful framework with which to
organize thoughts about energy and nutrient transfer. However, real-
world situations are usually far more complex. First, the notion of
discrete trophic levels is questionable. Where would an omnivore fit into
this scheme? Where should we place a carnivore that feeds on other
carnivores as well as on herbivores? Second, the notion of two distinct
pathways of energy and nutrient flow does not stand up to close inspection. Detritus-feeders may well be consumed by organisms in the grazing
food chain. Insectivorous birds and mammals are unlikely to discriminate
between insects on the basis of whether they belong to the grazing or
detritus food chain; and carnivores such as the big cats of Africa,
associated usually with the grazing food chain, may well occasionally
consume a mammal, such as the aardvark, that lives partly on detritus-
eating insects.

In aquatic ecosystems, there are close links between the grazing and detritus pathways as exemplified by the food chain in Fig. 9.5. Here, the photosynthetic algae continually release organic substances that are then assimilated by heterotrophic bacteria. These bacteria may then be consumed by protozoans, which in turn are a source of food for the tiny crustaceans of the zooplankton. However, since these crustaceans also consume the photosynthetic algae of the phytoplankton, they occupy a distinctly ambiguous position with respect to the grazing and detritus food chains.

A not dissimilar situation occurs with ruminants. Recall that microbial residents of the enlarged stomach of these animals break down cellulose into chemical components that are small enough to cross the gut wall. So the cellulose is used first, not by the grazing animal (cow, camel, llama etc.), but by the microbes that live in its digestive tract. Furthermore, the plant material is no longer living by the time it passes into the rumen. So, like the bacteria of the open water mentioned above, the gut microbes could be regarded as intermediaries between the detritus and the grazing food chain. And the 'herbivore' could be regarded as a functional detritivore!

Of course, the diagrammatic representation of feeding relationships requires neither distinct grazing and detritus pathways nor discrete trophic levels. However, as we shall see later, such a framework is a very useful one for analyzing food chains quantitatively and as a basis for comparing ecosystems.

Quantifying energy flow

We have seriously questioned the idea that each organism can be assigned, on the basis of its diet, to a particular trophic level in either the grazing or the detritus food chain. So, is this framework still a valid one for examining energy flow through ecological communities? The answer is a qualified 'yes', qualified for the reasons we have given above, and no doubt the qualifications needed will vary between situations. The fact is that in attempting to simulate the complexities of the real world we frequently need to make some concessions, and simply acknowledge that the model we construct is not a perfect representation of the real thing.

If such a model of energy flow is acceptable, it is a useful one, allowing numerical values to be derived for characterizing ecosystems of different sorts and therefore facilitating comparison between them. A variety of values can be calculated: the ratio of productivity to biomass at each trophic level; the ratio of biomass between trophic levels; the ratio of productivity between trophic levels; the level of exploitation of one trophic level by organisms in the next trophic level and the efficiency of energy transfer through the food chain.

The simple flow diagram in Fig. 9.6 shows how energy flow can be quantified. The grazing food chain represented here is a hypothetical one, but the values are realistic and from them some basic principles

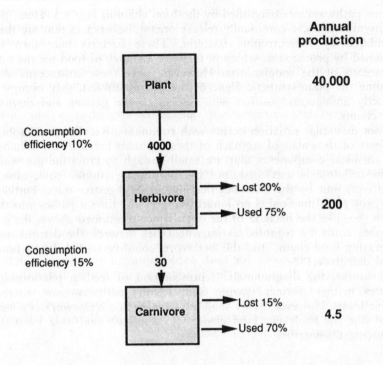

Fig. 9.6 Quantifying energy flow through an idealized grazing food chain. The three boxes represent the plants, a herbivore and a carnivore respectively. The values, which are in kilojoules $m^{-2} yr^{-1}$, are hypothetical, but realistic. The primary production value represents a fairly productive terrestrial situation. Consumption efficiency is the proportion of the annual production at one trophic level consumed by the animal at the next trophic level. 'Uses' are the respiratory activities, 'losses' are mainly undigested food.

emerge. For simplicity we envisage one herbivore species and one carnivore, but the same principles apply if we are dealing with several populations of herbivore and carnivore.

Follow the numbers through from primary (plant) production to the production of the carnivore. Exploitation efficiencies and respiratory and excretory losses are shown. (Eliminated substances are not necessarily lost to the whole community of course as the energy and nutrient content may be passed to detritus-feeders and indeed may re-enter the grazing food chain.) Note that the productivity decreases markedly from one trophic level to the next. Why this is so should be clear from the principles introduced earlier. It is partly because secondary producers do not normally exploit their food source with anything like 100 per cent efficiency, and it is partly because only a small proportion of the consumed energy ends up as new growth because of the losses (faeces, etc.), and uses (basal metabolism, activity etc.) as discussed in Chapter 8.

There is no satisfactory explanation as to why only a small proportion of available energy at one trophic level should normally be consumed by organisms in the next trophic level and there may be no unifying answer to this intriguing question. Whatever the explanation, a combination of low exploitation efficiency and low conversion efficiency by consumers results in a very large decline in productivity between one trophic level and the next. However, in some situations there may be a substantial input of energy and nutrients from outside the area of interest or, alternatively, a large transfer from the detritus food chain to the grazing food chain.

In general though, the decline in productivity from one trophic level to the next theoretically sets a limit to the number of trophic levels; this is simply because food becomes increasingly scarce with each additional trophic level. In fact, food chains in the real world (bearing in mind the problems of categorizing species in this way) usually have no more than three, four or five trophic levels. The ratio of productivities between trophic levels varies considerably, as we would expect from the range of values we have presented for both exploitation of primary production and the efficiency of food use for productive purposes by heterotrophic organisms. The ratio of productivity to biomass also varies. It is much lower in terrestrial communities than in aquatic situations simply because the biomass of the primary producers on land is usually many orders of magnitude greater than in open water.

The influence of Raymond Lindeman

The theoretical background to the study of energy flow between trophic levels was presented in an important paper, published in 1942, called 'The trophic-dynamic aspect of ecology'. Its author was Raymond Lindeman, a young research ecologist then working at Yale University, whose ideas had developed while working on a doctorate in aquatic ecology at the University of Minnesota. He was by no means the first ecologist to think about the organization of communities in terms of food chains, but his seminal paper laid the foundation for the development of quantitative energy-flow studies that took place during succeeding decades. Interestingly, Lindeman had quite a struggle to get this paper published, although it has become one of the benchmark papers in ecology. Sadly, Lindeman died while still in his twenties, before his paper became widely appreciated, so he was not to know how much influence his ideas would have on ecological theory.

Animals or plants in the human diet

The large decline in productivity that occurs between primary and secondary producers has led some people to call for less use of animal products in order to make more food available to the human population.

Although, superficially, it is an attractive argument, in many situations there are considerable advantages to using animals as a source of food, whatever ethical objections might be held against such practice. Harvesting the primary producers of oceans and lakes, for example, would not be a very practical proposition. (In fact most of the aquatic harvest is of carnivorous fish.) Also, we have seen that ruminant animals, because of the microbes in their digestive tracts, utilize plant products not available to humans. Living animals also effectively store food for long periods whereas harvested plant products have a limited life unless maintained under special conditions. In addition, animals provide food in situations where cultivation of soils for crop production would be ill-advised because of the risks of erosion.

The value of food chain studies

The study of feeding relationships and the movement of energy and materials through communities is not just a scientific curiosity. We are all completely dependent on the productivity from ecosystems of varying degrees of naturalness. Understanding this aspect of ecosystem function is important for the efficient management of biological resources. In addition, food chain studies provide a clearer picture of the behaviour of poisonous substances in the environment, a matter of considerable significance for ecological communities and for public health.

The energy 'metabolism' of whole communities

The focus so far in this chapter has been the transfer of energy and matter *within* communities. Of interest also is the balance between energy and carbon fixed in photosynthesis and that lost in respiration (and fermentation) from the whole community. The behaviour of the community – and indeed the whole biosphere – in this respect is determined by the collective activities of all the organisms present. If the amount of energy (or carbon) fixed in gross photosynthesis exceeds that lost in respiration during a particular time interval, chemical energy and organic matter accumulate. Conversely, if 'community respiration' exceeds gross photosynthesis, there must be a decline in the amount of organic matter and energy stored. The gain or loss may be represented by a change in either the amount of living biomass or dead organic matter, or both. Present concerns about rising levels of atmospheric carbon dioxide render studies of carbon and energy balances for the Earth's biota highly relevant, but we defer further discussion on this issue until the next chapter.

10 *Biogeochemical circulation*

The term biogeochemical circulation refers essentially to the reciprocal movement of chemical entities between organisms and the physical environment. The study of biogeochemical cycles encompasses consideration of the forms in which chemical elements are stored and transferred, the processes involved in their transfer and the chemical transformations that occur as elements move through the biosphere.

Biogeochemical circulation is an important theme in ecology as well as other areas of environmental study. As pointed out in Chapter 2, it is not only nutrients that are taken up by organisms and transferred along food chains, but other elements as well, including some such as mercury, lead and cadmium that are highly toxic. In addition, some nutrients, nitrogen for example, exist in molecular forms which are considered to be pollutants.

The point made at the beginning of the previous chapter concerning the close links between the movement of energy and nutrients is again emphasized. The concept of the food chain is just as appropriate for chemical elements as for energy. However, there is one fundamental difference between the movement of energy and chemical elements through the biosphere. Once chemical energy is used for work in a living cell it is transformed to heat, whereas atoms of the various elements are present in finite supply and theoretically can be used over and over again. So it is not correct to talk about 'energy circulation' whereas the expression 'nutrient circulation' is quite legitimate.

In this chapter we first consider some basic principles and processes relevant to biogeochemical circulation, then look at three contrasting studies and finally discuss the circulation of two key elements, nitrogen and carbon.

Some basic principles

Biogeochemical circulation can be studied over a variety of spatial scales. On the one hand, interest may centre on quite local movements, perhaps within a pond or part of a forest; at the other extreme, the objective may be to compile a global budget for a particular element. Although the detailed results generated by a study within one particular area will be specific to that area, general principles may emerge that can be applied to analogous situations elsewhere.

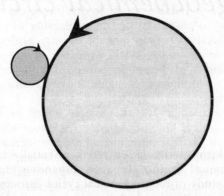

Fig. 10.1 Biogeochemical circulation. The smaller of the two circles represents short-term movements between the biota and the local physical environment; the larger circle represents a long time interval during which atoms are not biologically available.

In considering the circulation of chemical elements, it is useful to distinguish between short-term, local movements, and those that operate over very much longer time-scales, during which individual atoms may be removed from circulation for millions of years. This distinction is represented simply by the two coupled circles in Fig. 10.1. To exemplify, consider a single atom, of potassium say, that is taken up by a growing plant in the spring, incorporated into leaf tissue and later returned to the soil in the autumn after the leaf falls and is decomposed. Such movements between plant and soil, represented by the smaller of the two circles, may continue for several years before the potassium atom is leached from the soil, carried in solution and deposited on the bed of a lake. If accumulating sediment covered our atom it would be removed from circulation. It could take millions of years and a sequence of events, including the filling of the lake with sediment, uplifting of the land surface and subsequent erosion, to bring the atom back into biological circulation. Such a long-term sequence of events is represented by the larger of the two circles. We can regard all atoms of the nutrient elements as having alternated between periods of rapid circulation and very long periods when they have been removed from circulation, although the different elements vary considerably in the time-scales involved.

Biological availability of elements

As pointed out in Chapter 2, each element contributes to a variety of chemical substances. The nature of the substance has an important bearing on an element's biological availability, i.e. the extent to which it can be taken up by organisms. This means in practice that a measure of the total amount of a particular element, in the rooting zone of soil or in water, may not be a reliable guide to its biological availablity. This applies particularly to phosphorus. The availability to plants of this key element is closely related to soil acidity. In very acid soils (i.e. of low pH), phosphorus tends to be poorly available. It becomes increasingly available, as the acidity of the soil is reduced, but in very alkaline conditions (high pH), phosphorus is again largely insoluble and poorly available for plants.

The biological availability of aluminium, which in solution is quite toxic, is also pH related. One of the concerns about soil acidification as a result of 'acid rain' is that aluminium becomes increasingly soluble as acidity increases. Also, soluble aluminium may be leached from soils and enter bodies of water. Aluminium poses a particular problem for fish because it impedes their capacity for oxygen uptake. This is one of the main reasons for the decline in fish populations associated with the acidification of waters.

Nutrients in water

Within bodies of water, apart from those that are very shallow, the vertical distribution of each nutrient has a major influence on its biological availability. The reason why this is so should be clear from our discussion of aquatic primary production in the previous chapter. You will recall that photosynthesis can occur in water only to the depth at which light is adequate for this process, i.e. in the region known as the euphotic zone. However, organic remains tend to sink through the water column, so that decomposition and release of nutrients may occur in large part below the euphotic zone. To be taken up by the photosynthetic algae, nutrients must therefore be physically transferred upwards. Now, annual primary production in marine and freshwater ecosystems is determined principally by the availability of certain key nutrients. It is no coincidence, therefore, that aquatic productivities tend to be relatively high where conditions are conducive to vertical mixing of water, but relatively low where little vertical mixing occurs.

Vertical mixing of water can be promoted in a number of ways. Wind activity generates turbulence, and hence stimulates mixing, particularly in near surface waters. Wind activity can also indirectly promote the upward movement of water, a movement known as *upwelling* (Fig. 10.2a). Upwelling occurs where surface waters, moved away from continental land masses by prevailing off-shore winds, are replaced by nutrient-rich water from below. Such upward movements of nutrient-rich water, which occur in parts of the eastern Atlantic off West Africa and in the Pacific Ocean off South and North America, account for the high biological productivities characteristic of these areas.

Gradients of water density (i.e. weight per unit volume) in the water column also have a major influence on vertical mixing. The density of fresh-water is determined by temperature; maximum density occurs at 4°C. The density of sea-water is determined by both temperature and

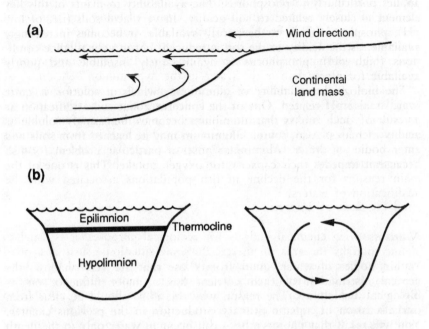

Fig. 10.2 Vertical mixing in water. (a) Upwelling: surface water, moved by off-shore winds, is replaced by nutrient-rich water from depth. (b) Density effects: in the left-hand diagram, warm, low-density water lies above cooler, higher-density water and little vertical mixing occurs. In the right-hand diagram, seasonal cooling has increased surface water density leading to an overturn of water.

salinity: the lower the temperature (at a given salinity) the higher the density, and the greater the salinity (at a given temperature) the higher the density. Where currents of different density converge, as in sub-Arctic latitudes in the Atlantic and Pacific Oceans and around the Antarctic continent, the cooler, denser water tends to sink, thus promoting vertical mixing. Nutrients are thus returned to, or retained within, the euphotic zone and in consequence biological productivity is relatively high.

The extent to which the temperature of surface water changes over the year also has a large influence on mixing, and hence on nutrient transfer. In tropical zones, where seasonal temperature changes are small, waters may be almost permanently layered, or *stratified* (Fig. 10.2b). A layer of warm, low-density water, the *epilimnion*, lies above a denser, cooler body of water, the *hypolimnion*. The two layers are separated by a comparatively narrow band of water, the *thermocline*, in which the temperature gradient is steep. Where conditions are likely to lead to almost permanent stratification, mixing due to wind activity or upwelling is critical for nutrient transfer.

Stratification is also characteristic of waters in temperate and higher latitudes during the summer months, but in these zones the autumnal cooling of surface water results in an 'overturn' of water and transfer of nutrients (Fig. 10.2b). Fresh-water bodies in seasonal climates typically undergo a second overturn in the spring. This occurs when the temperature of the warming surface waters is closer to 4°C (the temperature of maximum density) than is the water in the hypolimnion beneath.

Seasonal overturns and wind-induced mixing also promote the downward transfer of oxygen. Oxygen replenishment is particularly important in waters which receive organic effluents or have high nutrient loadings and thus high productivities. Oxygen demand is high in such situations on account of the activities of decomposer organisms and its rate of consumption may exceed the rate at which it can be replenished.

Studying chemical budgets

When studying chemical budgets for ecological purposes, it is usual to define spatially the area of interest. In some situations, such as a pond or a river catchment, a boundary may be fairly well-defined by natural features. In other cases drawing boundaries is a more subjective process. By delimiting an area of study, however arbitrarily, it is possible to address questions concerning the relative 'openness' of ecological systems with respect to element circulation. In this context, 'open' ecosystems are those which import and export large amounts of nutrients while 'closed' systems are those characterized by small throughputs of nutrients. Of course, these terms are only relative; no part of the biosphere is either entirely closed or entirely open with respect to chemical elements. However, a stretch of stream could be reasonably described as a relatively open system due to the continuous throughput of nutrients: undisturbed tropical rain forest, in contrast, would normally be considered a relatively closed system because rates of chemical 'leakage' are fairly low. We shall consider inputs and outputs in more detail a little later.

Stores of chemical elements

In studies of ecological chemical budgets it is customary also to identify various 'compartments' or 'pools' of elements within the area of interest, and to determine their relative sizes and the movements between them. A very simple compartment model is shown in Fig. 10.3 as a framework for further discussion on this topic. In fact, both the physical environment and dead organic matter boxes could be subdivided to represent the chemically available and unavailable fractions that were referred to earlier, but this model will suffice for now. The distribution of chemical elements between the three compartments shown varies considerably, and in characteristic ways, according to the type of ecosystem. However, it

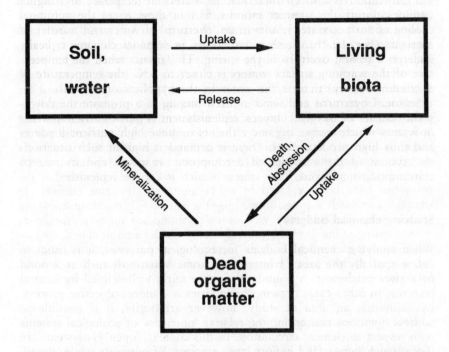

Fig. 10.3 A simple compartment model showing the main stores of mineral nutrients and the pathways of transfer between them. Note that nutrients within the physical environment and dead organic matter are not necessarily in a biologically available form. The thicker arrows represent the major routes of transfer.

should not be assumed that all elements behave in exactly the same way in the same area. If we confined our study to phosphorus, for example, we should not assume that all other elements behaved similarly.

In most situations, the physical environment accounts for a high proportion of each mineral element, although the proportion varies between ecosystem types and between elements. For example, the nutrient ratio between soil and vegetation tends to be higher in temperate forests than in humid tropical forests. The distribution of nutrients between the living and dead organic compartments also varies. In temperate and high-latitude forests, a greater proportion of the nutrient capital is held in the litter at the soil surface than is normally the case in tropical forests where nutrients are rapidly returned to the living biomass.

Within the living biota, it is customary to identify a number of subcompartments, as our previous discussion of food chains would suggest. On land, the vegetation usually accounts for most of the biotic nutrient capital, but in aquatic ecosystems, where the primary producers are tiny and very short-lived, the animals may account for a much greater proportion of the nutrients held in the biota.

In studies of terrestrial ecosystems, the vegetation is often subdivided into various components, by growth form perhaps (trees, shrubs, herbs, etc.), by species, and also by tissue types (leaves, stems, roots, etc.). Chemical analyses of such components reveal how elements are distributed at an instant in time, but are not particularly informative about the dynamics of element circulation. For example, in non-seasonal environments the distribution of mineral elements among plant tissues may not appear to change very much over the year. Such a finding should not be taken as evidence that no nutrient movement has occurred; it is likely that the atoms of the various elements, present in leaves when the vegetation was first sampled, have since been *translocated* elsewhere in the plant but have been replaced by other atoms of the same element. In environments with a pronounced growing season, chemical analysis and growth measurements may well reveal pronounced nutrient uptake at certain times of the year. In general though, a detailed understanding of the dynamics of element circulation requires special procedures, such as the use of radioisotopes whose movement through the system can be followed.

Chemical movements within ecosystems

The arrows in Fig. 10.3 indicate the principal directions of movement for mineral elements in ecosystems. On land, uptake by plants is the major routeway of chemical transfer from the physical environment to the biota. Some qualification is necessary here though because fungi, associated with plant roots, seemingly always account for a significant amount of mineral nutrient transfer from soil to vegetation. Such associations between plant roots and fungi, termed *mycorrhizae*, are very important both for biogeochemical cycling and for the nutrition of most plants. Essentially there are two types of mycorrhizal association and they involve different fungal species. In one, termed *ecto*mycorrhizae, fungi are restricted to the root surface, whereas in the other, *endo*mycorrhizae, the hyphae penetrate the outer cells of the root. In both types, hyphae extend some distance from the root and considerably enhance the capacity of the host plant to procure nutrients. (In return for supplying mineral nutrients to the plant, the heterotrophic fungus receives a supply of organic assimilates.)

In moist tropical environments it has been shown that mycorrhizal fungi can transfer nutrient elements directly from decaying organic matter to plant roots, thus the thin arrow from the dead organic matter pool to the living biota in Fig. 10.3. Just how important this routeway of nutrient transfer is remains unknown, but such a feature should enhance nutrient conservation in the wet tropics where the potential for leaching is high.

In open water, nutrient uptake is brought about principally by the microscopic algae that are usually the dominant members of the phyto-plankton. In near-shore environments, larger plants (aquatic *macro-phytes*) such as the 'seaweeds' and the seagrasses of shallow marine environments, and the various floating and rooting plants of freshwater environments, are often important for nutrient uptake. The rate at which nutrients are transferred between the environment and the biota is usually related to seasonal growth patterns of the dominant primary producers. During times of the year when little growth of these organisms occurs, nutrient uptake may be very low or cease altogether.

Death of organisms and the shedding of tissues brings about a transfer of elements from the living biota to the dead organic matter pool (Fig. 10.3). Leaf shedding is a particularly important mechanism for nutrient transfer in forests. In deciduous forests, leaf shedding is usually concen-trated into a short period at the onset of either the cold or the dry season. In evergreen forests, such as the coniferous forests of the northern hemisphere and the broad-leaved forests of the permanently-wet tropics, leaf shedding may be less concentrated. It seems that the withdrawal of nutrients from leaves prior to shedding may be a quite common feature of deciduous trees, thereby conserving nutrients for the trees.

Nutrients can be transferred directly from living organisms to the physical environment, as when plant leaves are leached by rainwater and metabolic by-products are excreted by animals. However, by far the most important way by which mineral nutrients are returned to the physical environment is through the decomposition of dead organic matter and the associated mineralization of chemical elements. The rate of mineralization is controlled by decomposer organisms, whose activities, as discussed in Chapter 9, are determined largely by environmental conditions.

Inputs and outputs of mineral elements

Earlier in this chapter we made the point that some types of ecological system are characteristically more open than others with respect to the movements of chemical elements. Imports and exports are high in some situations and relatively low in others. The purpose of this rather long section is to provide a framework for the study of mineral element transfer, emphasizing the processes which result in gains to and losses from prescribed areas of land and water. Of course, we should bear in mind earlier comments concerning the spatially continuous nature of the environment, so that prescribing areas is a rather arbitrary process. We should also not forget an earlier comment about the finite nature of chemical elements; so losses from one area mean gains elsewhere, whilst gains in one area must mean losses somewhere else.

A framework scheme with which to consider inputs and outputs is shown in Fig. 10.4. Of course, there is no suggestion that for any given area all potential inputs and outputs are worthy of consideration;

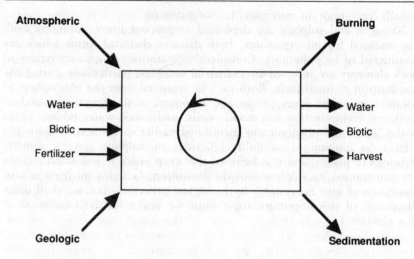

Fig. 10.4 General scheme for considering inputs and outputs of chemical elements for ecological systems. (Biological nitrogen fixation and denitrification are omitted.) Usually only some of the various inputs and outputs will be important.

indeed, terrestrial and aquatic ecosystems are deliberately treated together here even though they differ in some very obvious ways.

Atmospheric inputs

The atmosphere is a potentially important source of chemical elements for ecological systems. From the atmosphere, elements may be introduced in solution in rain and snow, or in aerosol form, and deposited 'dry' between precipitation events. Some elements, notably nitrogen and sulphur, may be introduced in gaseous form. Certain vegetation types, particularly forests, effectively trap elements held within mist and cloud. Where such conditions are common, as in some mountain and coastal locations, this input can be significant. When aerial inputs are assumed to be represented only by what is found within a standard rain gauge (known as bulk precipitation) this last input is not accounted for, so the aerial input can be seriously underestimated. Aerial inputs of some elements may be significant in coastal habitats because of the proximity of sea-water with its complement of cations (sodium, potassium, magnesium and calcium) and anions (chloride and sulphate).

Atmospheric inputs are extremely important for the nutrient economy of communities isolated from other sources. In particular, peat may accumulate to such an extent that it is no longer supplied by surface water and the vegetation growing on it is isolated from the mineral soil beneath. In such a situation, the only nutrient additions are from the atmosphere. Peat bogs of this sort, described as *ombrogenous*, are usually very poor in nutrients, i.e. *oligotrophic*.

Nitrogen and sulphur are deposited in gaseous form. Although both are essential for all organisms, both exist in chemical forms which are considered to be pollutants. Considerable quantities of gaseous oxides of both elements are released by industrial societies, particularly during the combustion of fossil fuels. Both may be removed from the atmosphere in solution in acidic form (nitric and sulphuric acid respectively) and in sufficient concentration can acidify soils, and hence water bodies. (This is the 'acid rain' phenomenon mentioned earlier in connection with pH effects on aluminium solubility.) Oxides of sulphur can be directly injurious to plants, and it is believed that crop yields in industrial regions are constrained by sulphur dioxide deposition. Gaseous nitrogen is also incorporated into ecosystems by biological processes, but we shall defer discussion of this important topic until we deal with the circulation of this element.

Geologic inputs and outputs

Nutrient elements are made available for ecological systems during the weathering of rocks and minerals. In fact, the weathering of such materials, with the release of nutrients, almost defines the process of soil formation. The geologic input varies considerably, depending on the nature of the parent material and prevailing environmental conditions. Soils formed on siliceous parent materials, such as granite, tend to have low concentrations of biologically essential elements. Soils formed from calcareous parent materials are potentially more fertile, although nutrients released by weathering are subject to losses by leaching processes in humid climates. Rates of mineral weathering vary according to climate, tending to be higher in warmer, humid conditions than in colder, dry environments.

The geologic component is often ignored in short-term biogeochemical studies, partly because obtaining reliable values is awkward, but partly because the contribution from this source is usually considered to be negligible. This is in fact one of the least well-quantified aspects of biogeochemical cycling and closer examination could show that its importance has been previously underestimated. It is also complicated by short-term changes in the biological availability of individual elements, as we discussed earlier for phosphorus.

In some ways analogous to the parent materials of soils are the sediments of the sea- and lake-floor. From here elements may be introduced, at rates determined by the nature of the substrate, local environmental conditions and biological activity. Minerals are also released into the oceans as a result of eruptive geological processes. We have referred earlier to the submarine hot-springs associated with mid-oceanic ridges where hydrogen sulphide is emitted and used by chemosynthetic bacteria. These locations are also the source of other minerals and such emissions seemingly make a significant contribution to the mineral budget of the oceans. From aquatic ecosystems, mineral elements are lost, at least from short-term circulation, by sedimentation processes. Again, however, we should bear in mind that it is difficult to draw a distinction between the minerals that are biologically available and those that are unavailable.

Element transfer in water

Flowing and percolating waters carry chemical elements, either in solution or in association with particulate matter. The extent to which lakes, estuaries and seas are influenced by rivers discharging into them depends on the chemical characteristics of the river water, flow rates and residence time of water. Under natural conditions, the chemistry of flowing and percolating water is determined primarily by the geological characteristics of the catchment area. Waters passing over much-weathered, siliceous, parent materials, such as the ancient rocks of the Canadian Shield, naturally carry low concentrations of elements, but rivers and streams passing over softer, more easily weathered materials can carry considerable chemical loads.

Of course, human activities within the drainage basin have a marked influence on the chemical characteristics of river water. Nitrogenous fertilizer, phosphorus-containing detergents and sewage pose particular problems because aquatic productivity is closely tied to the availability of key nutrients. Deforestation, particularly on steep slopes in humid environments, is a major cause of nutrient, and also particulate matter, transfer from land to water. But human activities can also reduce nutrient transfer. The Aswan High Dam, constructed in Egypt during the 1960s, trapped sediment so efficiently that the nutrient load of the water reaching the Mediterranean declined markedly. As a result, biological productivity off-shore declined with serious consequences for the local fishery.

Mineral-rich waters may raise or maintain nutrient levels, and also the pH, of terrestrial environments. Such is the case with peaty areas that receive nutrients in solution. (Such peats are termed *fens* to distinguish them from acidic, mineral-poor peats, usually referred to as *bogs*.) Within otherwise nutrient-poor situations, as occur over much of the British uplands, the emergence of spring and stream-water may raise fertility locally and support plant species which are not characteristic of the surrounding nutrient-poor area. In humid environments, the predominantly downward movement of water through soil provides a potentially important mechanism for nutrient loss. Elements vary considerably in the extent to which they are vulnerable to loss by leaching. Nitrogen is relatively susceptible to leaching losses while phosphorus is in general a much less mobile element.

Soils, too, vary in their capacity to hold elements, particularly the cationic elements such as potassium, calcium and magnesium. Cations are adsorbed on to negatively-charged surfaces of clays and organic matter. Soils rich in clay minerals and/or organic matter typically have a relatively high *cation exchange capacity*, and in consequence are generally less prone to leaching losses than are sandy soils which are chemically much less active.

Biological transfers

In some situations, mineral elements are transferred in ecologically significant amounts by the movements of living organisms. The movements of geese, gulls and hippopotami, which feed on land during part of the day, and spend part of the day in water, have all been shown to bring about nutrient transfers from land to water. By this means, large populations of gulls have significantly raised the fertility of at least one of the shallow lakes that make up the Norfolk Broads in eastern England with a consequent decline in its ecological and amenity value. Inputs of mineral elements in shed plant material, particularly leaves, but sometimes pollen as well, can also make a significant contribution to the nutrient economy of streams and ponds.

Fire

Fire is an important ecological agent over a significant proportion of the Earth's land surface. Fires are caused naturally by lightning, with volcanic activity playing a minor, local role. But fire is also used as a tool in vegetation management, and of course many large-scale burns are started accidentally. As a general rule, burning of vegetation tends to reduce the nutrient capital in an area. Losses occur during the fire itself as minerals are converted to gaseous form (a process known as *volatilization*), or removed as particulate matter in the smoke. The magnitude of the loss depends largely on the temperature of the burn, and mineral elements differ in their susceptibility to volatilization. Nitrogen and sulphur are lost in gaseous form at relatively low temperatures; phosphorus is lost less readily by this means followed by the cationic elements.

However, burning also predisposes nutrients to loss through leaching and erosion. When vegetation and litter are burned there is a rapid transfer of mineral elements from the organic pools to the physical environment. Nutrients in the ash are vulnerable to loss by leaching and wind erosion, particularly as the destruction of a protective vegetation cover usually leads to higher rates of rainfall infiltration and higher wind speeds at the ground surface.

Fertilization and harvest

Nutrient inputs as fertilizers and outputs as harvested products may be

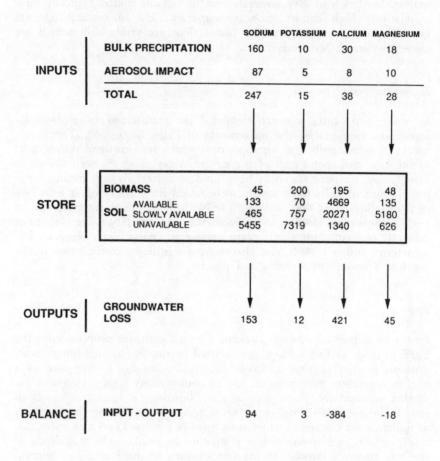

		SODIUM	POTASSIUM	CALCIUM	MAGNESIUM
INPUTS	**BULK PRECIPITATION**	160	10	30	18
	AEROSOL IMPACT	87	5	8	10
	TOTAL	247	15	38	28
STORE	**BIOMASS**	45	200	195	48
	SOIL AVAILABLE	133	70	4669	135
	SLOWLY AVAILABLE	465	757	20271	5180
	UNAVAILABLE	5455	7319	1340	626
OUTPUTS	**GROUNDWATER LOSS**	153	12	421	45
BALANCE	**INPUT - OUTPUT**	94	3	-384	-18

Fig. 10.5 Measurements of aerial inputs and groundwater losses (kg ha^{-1} yr^{-1}) and stores (kg ha^{-1}) of some cations for part of the Ainsdale sand dune system, north-west England. Note the determination of both bulk precipitation and aerosol impact and also the partitioning of the soil nutrient store on the basis of availability.

very important for agricultural ecosystems. Such systems are therefore quite open with respect to nutrient transfer. Usually it is nitrogen which is applied in the greatest amounts, but large quantities of fertilizer phosphorus and potassium are also frequently added. Mining minerals for fertilizer manufacture, as in the case of phosphorus, potassium and calcium, has the effect of short-circuiting the natural long-term mineral cycles that we referred to near the beginning of this chapter. Most of the nitrogen fertilizer applied to soil is manufactured following the industrial fixation of atmospheric nitrogen, a process that is now an important component of the global nitrogen cycle.

The quantities of the various nutrients exported will vary according to the nature and the amount of material that is harvested and removed. In addition, however, agricultural practices have a less direct effect on nutrient losses. Soil that is left with no protective cover of vegetation is inherently vulnerable to erosion and leaching losses. The magnitude of such losses will depend on a number of factors, most notably topography, climate and soil characteristics.

Examples of element circulation studies

Here we briefly introduce three very different nutrient circulation studies to illustrate the wide range of work encompassed by this field of enquiry. First, in Fig. 10.5 some results are shown from a study of major cations in a sand dune system in north-west England. The location of this area suggested that the marine influence was likely to be strong, and so special efforts were made to quantify the inputs in aerosol form as well as in solution in rainwater. On the output side the major routeway was considered to be groundwater discharge. Notice also how the total complement of each cation within the dune system has been partitioned into 'available', 'slowly available' and 'unavailable' fractions.

The second example is a long-term study, initiated during the 1960s, in forested catchments in the state of New Hampshire, north-eastern United States. A major aim was to determine the effects of vegetation removal on the nutrient budgets in such environments. The most important medium for nutrient loss was considered to be water, so streams emanating from the watersheds were dammed, water losses measured, and samples analysed for their nutrient content. The long-term nature of this study and the large amount of effort devoted to it have made this study a 'classic'. Notice in Fig. 10.6 that the export of nutrients is rapidly increased immediately following clearfelling, and also that the high rates of loss were maintained as long as vegetation regrowth was suppressed. The enhanced loss of nutrients from the deforested watershed is attributable partly to the increased amounts of rainfall that reached and percolated through the soil, but partly also to the impaired capacity of the biota to take up nutrients from the soil, which of course predisposed nutrients to loss by leaching.

The third example (Fig. 10.7) is rather unusual. It shows how the principles of nutrient circulation can be applied nationally. The aim of this study was to identify the main inputs and outputs of phosphorus for the United Kingdom, and to determine the major internal stores and transfer routes for this element. Phosphorus is of course a key nutrient; large amounts are applied annually to agricultural crops and it is also used for industrial and domestic purposes. Phosphorus is a valuable and expensive resource, but as we have previously mentioned, its discharge into water bodies can have significant ecological consequences. Such studies may suggest ways in which phosphorus could be used more efficiently with reduced discharge and waste.

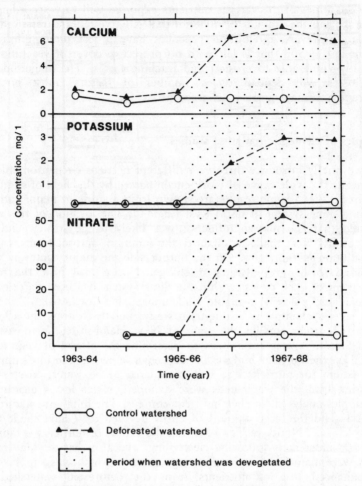

Fig. 10.6   The effect of deforestation and supression of regrowth on the release of some
            nutrients in streamwater from a catchment at Hubbard Brook, New
            Hampshire. Notice how forest clearing precipitates the rapid and large-scale
            export of nutrients.

The biological nitrogen cycle

The behaviour of nitrogen is unique among the biologically important
elements. It exists as part of a large number of chemical entities and its
circulation in the biosphere involves several chemical and biochemical
transformations. Some of the substances in which it occurs, particularly
the oxides of nitrogen, are present in sufficient amounts in the environment to be considered serious pollutants.

   An outline scheme for the biological nitrogen cycle is shown in Fig.
10.8. Nitrogen accounts for almost eighty per cent by volume of the

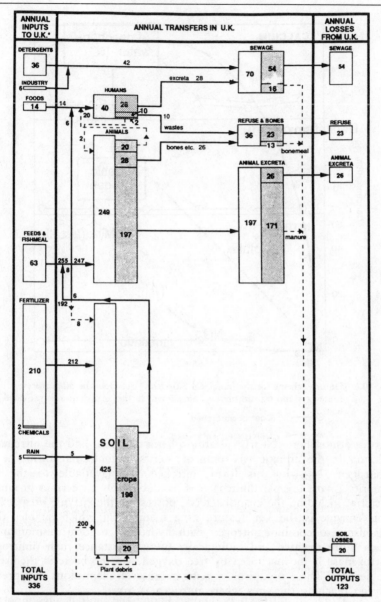

1. All figures in thousand tonnes P, rounded to nearest thousand.
2. Data based on 1973 except where otherwise stated in text.
3. * With the exception of rain, the annual inputs refer to imports of phosphorus.

Inputs   Outputs

Fig. 10.7 An annual phosphorus budget for the United Kingdom. Note the relative magnitude of the various inputs and outputs and the movement of phosphorus within the UK.

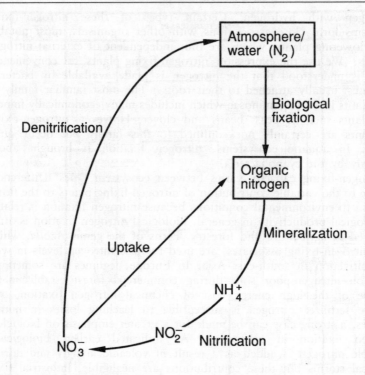

Fig. 10.8 Outline scheme for the biological nitrogen cycle. Only the main stores, routeways and transformations are shown; further explanation is provided in text.

Earth's atmosphere. Nearly all atmospheric nitrogen, and the nitrogen in solution in the aquatic environment, exists in molecular form, i.e. as dinitrogen (N_2). In this form, nitrogen is unavailable to the vast majority of organisms: dinitrogen is available only to certain groups of bacteria, including the cyanobacteria, possessing the enzyme *nitrogenase*. This enzyme is the key feature of a metabolism which enables these organisms to combine nitrogen with hydrogen to form ammonium, a much more reactive and biologically useful substance than dinitrogen. Nitrogenase is incapacitated by free oxygen, yet the bacteria themselves are aerobic. These organisms survive because of adaptations that prevent oxygen reaching the sites where nitrogen is 'fixed'.

Biological nitrogen fixation is carried out at normal temperatures and pressures, unlike the industrial fixation process which requires a temperature of 700°C, pressures 100 times greater than atmospheric pressure and a metallic catalyst. Nitrogen fixation must have occurred early in the history of life. Previously, organisms would have been dependent on the limited amount of combined nitrogen produced during volcanic activity and electrical storms.

So it is only prokaryotic organisms that have the capacity to combine dinitrogen with hydrogen. Certain types of these nitrogen-fixing organisms form close associations with other organisms, most notably some flowering plants, which are thus independent of external nitrogen supplies. (We use the expression 'nitrogen-fixing plants' for convenience, but it is understood that the nitrogen is made available by bacterial associates, usually attached to their roots.) The most familiar family of such plants is the Leguminosae which includes many economically important plants such as peas, beans and clover. However, nitrogen-fixing symbionts are certainly not confined to this family but occur quite widely. In aquatic ecosystems, nitrogen fixation is brought about primarily by the cyanobacteria.

Nitrogen-fixing capacity varies between ecosystem types. Differences are due to the variable contribution of nitrogen-fixing plants to the flora, but also to environmental conditions because nitrogen fixation is related to biological productivity in general. Biological nitrogen fixation is often exploited in agriculture and forestry. Ferns of the genus *Azolla*, which have nitrogen-fixing associates, are used to raise nitrogen levels in wet-rice cultivation in south-east Asia. In Europe, legumes are sometimes sown on nitrogen-poor soils during commercial forest establishment. Because of the high energy cost of chemical nitrogen fixation, and because fertilizer nitrogen is susceptible to leaching losses in humid climates, a strong case can be made for a greater emphasis on biological nitrogen fixation in agriculture. As indicated earlier, biologically available nitrogen is added as a result of volcanic activity and during electrical storms but these contributions are negligible. Industrial fixation, however, primarily for fertilizer manufacture, is a very major component of the whole nitrogen cycle.

Following its combination with hydrogen, nitrogen is incorporated into a range of substances, notably proteins and nucleic acids. In our scheme (Fig. 10.8), the organic pool is represented by a single box, but of course this could be partitioned into living and dead compartments, into plants, animals and microorganisms, and into various species and tissue types as we suggested earlier when considering element circulation in general.

Typically, well over ninety per cent of the nitrogen in soil is in organic form, and most is in organic remains rather than in living tissue. In open water, however, only about half the combined nitrogen is in organic form, the remainder being held in inorganic substances. (Recall that in water most of the *total* nitrogen exists as gaseous dinitrogen.)

The concept of biological availability that we introduced near the start of this chapter is particularly relevant for nitrogen. Nitrogen within the dead organic pool is not available to most organisms. A few types of bacteria and fungi can utilize organic nitrogen, and in so doing they liberate ammonium ions. This is the mineralization step. In aquatic environments, the tiny animals comprising the zooplankton also make a significant contribution to the available nitrogen in the water by excreting ammonia.

Ammonium ions in soil and water-bodies can be assimilated by certain organisms, including some plants. More importantly, ammonium ions

are used chemosynthetically (see page 81) as an energy source by certain
types of bacteria which release nitrite (NO_2) ions in the process. Nitrite
ions are used chemosynthetically in turn by other types of bacteria and
these release nitrate (NO_3) ions. These two steps are referred to as
nitrification (they were used to exemplify chemosynthesis in Chapter 7).
The second stage of the nitrification process proceeds more readily than
the first in soil, so nitrite ions do not normally accumulate.

Nitrification is vital to the functioning of the biosphere because most
of the nitrogen taken up by the primary producers is as nitrate. However
nitrate ions are very mobile and easily leached from soil in drainage
water if not biologically assimilated. Once incorporated within a plant or
microbe, nitrogen atoms can pass along food chains like any other
element. Nitrogen returns to the dead organic pool in soil or water when
organisms die, when leaves fall or when urine and faeces are released. In
fact, urine and other excreted by-products of animal metabolism contain
nitrogen in abundance. Since nitrogen in this form is readily available,
grazing animals have the effect of speeding up the movement of nitrogen
in ecological systems.

Nitrification proceeds only if sufficient oxygen is available: under
anaerobic conditions some types of bacteria substitute nitrate ions for
oxygen in their energy metabolism and as a consequence gaseous
nitrogen is formed. This process, known as *denitrification*, returns
nitrogen to the atmosphere or to the water. Natural denitrification is
now supplemented considerably by the release of nitrogenous gases in
combustion processes, but under natural conditions the assumption is
made that for the Earth as a whole denitrification would more or less
balance nitrogen fixation.

The carbon cycle

It should be clear that carbon plays a unique role in the biosphere; the
chemistry of life is very largely the chemistry of carbon compounds. In
previous chapters we have considered the processes of photosynthesis and
respiration by which carbon is transferred between living organisms and
their environment. Here, the focus is on the global budget for this
element, particularly the major stores of carbon and pathways of movement between them. Much attention has been paid to global carbon
budgeting during the last two decades; some aspects have become better
understood but there are still considerable areas of uncertainty,
particularly regarding the amounts involved. A global carbon budget is
shown in Fig. 10.9. The values in this diagram were suggested about a
decade ago by the American ecologist George Woodwell. The range of
values for carbon in some compartments reflects the incomplete state of
knowledge and the room for error that exists in compiling a global
budget for this element.

Notice first that sediments account for well over ninety-nine per cent
of all the carbon on the planet. Most of the carbon in this compartment

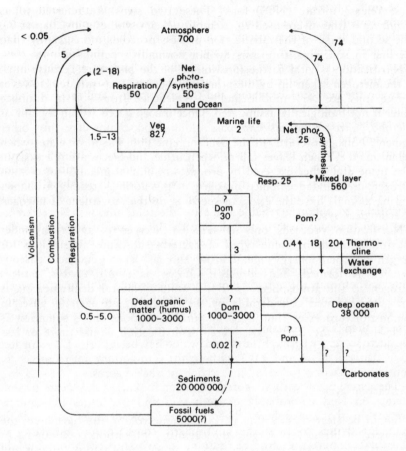

Fig. 10.9 The global carbon budget. Values shown are billions of tonnes; fluxes, per annum. Note there is considerable uncertainty concerning the magnitude of some of the carbon reservoirs and fluxes.

occurs in carbonate rocks, but a significant proportion exists in the hydrocarbons on which our fossil fuels are based. Under natural conditions the exchange of carbon between sediments and the water and atmosphere is quite slow; carbon released by rock weathering is eventually dissolved in the oceans and deposited on the ocean floor. However, the burning of fossil fuels on a large scale over the last hundred and thirty years or so has markedly enhanced the rate of carbon release from the sediment compartment. The use of carbonates for cement production also releases carbon from the sedimentary compartment.

Second, observe that most of the carbon outside the sedimentary compartment is in the oceans. By far the greatest amount of oceanic carbon is in the 'deep ocean' compartment, which is approximately equivalent to that part of the water column below the euphotic zone. Here, carbon has a very long residence time, exchanging only very slowly with that in the waters above. Carbon exchanges much more readily between the near-surface waters and the atmosphere, and there is a rapid transfer between the organic and inorganic pools in this zone due to photosynthesis and respiration.

Within the large reservoir of inorganic carbon in the oceans, around ninety per cent occurs as bicarbonate ions (HCO_3^-), around ten per cent as carbonate ions (CO_3^{2-}) and less than one per cent as carbon dioxide. Organic carbon in the oceans exists primarily as dissolved and particulate organic matter, but because of its highly dispersed nature it is difficult to obtain a reliable value for the total amount. No reference has been made to fresh-water, despite the large size of some lakes and river systems, because collectively they account for just a tiny fraction of the volume of water held in the oceans.

The terrestrial biota, including the wood of living trees, stores vastly more carbon than the aquatic biota, possibly around 300 times as much. This does not mean that the primary productivity ratio between land and the oceans is as great; it is just that, on average, land plants live so much longer than their functional equivalents in open water. The annual productivity ratio between land and water has been variously estimated to lie between 1.3:1 and 3.0:1, although it is important not to forget the considerable scope for error in arriving at such figures.

The amount of carbon stored in living biomass and dead organic matter on land is probably between two and five times the amount present in the atmosphere. There is thus considerable potential for human activities to raise the atmospheric concentration. Activities of particular importance are the burning of woody vegetation and the cultivation of soils. Oxidation of carbon occurs very rapidly when organic matter is burnt, but rather more slowly when soils are aerated during cultivation. Burning has a history extending back several millenia in some parts of the world, but attention now is focused on carbon fluxes resulting from the burning of forests in tropical zones.

Until about 1960, the total amount of carbon released as a result of forest burning and soil cultivation probably exceeded that due to fossil fuel combustion. However, the combustion of fossil fuel has continued to increase and is now believed to be the most important source of carbon. Again, though, it is important to emphasize that estimates vary widely on account of the difficulties of determining the magnitude of carbon release through forest burning.

The concentration of carbon dioxide in the atmosphere has continued to rise from the late nineteenth century, when values were probably

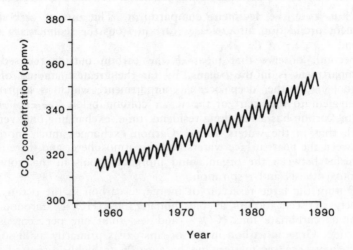

Fig. 10.10 Temporal trend in atmospheric carbon dioxide concentration as determined by measurements at Mauna Loa in the Hawaiian Islands. Note the fluctuations, assumed to be due to seasonal changes in the balance between photosynthesis and respiration on the northern continents.

around 290 ppm by volume. Reliable measurements of atmospheric carbon dioxide have been available from the mid-1950s when the concentration was about 315 ppm. Since then, the rise in carbon dioxide concentration has continued; the 1990 value is about 340 ppm and if the present rate of increase is maintained it is destined to reach 600 ppm by the middle of the twenty-first century. Notice in Fig 10.10 that carbon dioxide levels fluctuate seasonally. This is the net effect of seasonal changes in the balance between photosynthesis and respiration in the northern hemisphere, and it serves to emphasize the important influence that biological processes can have on the physical environment.

The observed increase in atmospheric carbon can account for only a proportion of that estimated to have been released as a consequence of human activities. It is convenient to assume that the residual amount has been absorbed by the oceans but, as pointed out earlier, carbon in this pool is relatively isolated from carbon in the waters above. It could be that considerable amounts of carbon have entered the oceans, gradually descended to the ocean floor and hence been lost from circulation. Alternatively, some of the unaccounted-for carbon may be stored in biomass having been assimilated in photosynthesis. It is extremely difficult to obtain reliable data for all the various components of the carbon cycle, so the fact is no one knows for sure.

There is no such doubt about the continued rise in the atmospheric concentration of carbon dioxide and other 'greenhouse' gases, notably methane. Atmospheric methane increased by over one per cent annually during the 1980s. As discussed earlier, this gas is generated under anaerobic conditions by certain types of bacteria and it is speculated that raised methane levels are due partly to an increase in cattle numbers but also in the area of wet-rice cultivation where anaerobic conditions prevail for part of the year.

It is widely believed that global warming, by inducing the expansion of oceanic water and some glacial melting, will result in a worldwide rise in sea levels, and also cause a shift in climatic zones. The really important questions now are; by how much, how soon, and can the trend be arrested?

Further reading

Begon, M., Harper, J.L. and Townsend, C. (1990) *Ecology – Organisms, populations and communities*, 2nd edn. Oxford, Blackwell.
(Considered by many to be the best of the substantive undergraduate ecology texts; authorititive, comprehensive and very well presented.)
Colinvaux, P. (1980) *Why Big Fierce Animals Are Rare*. London, Penguin Books.
(Intriguing title – adapted frequently for examination purposes – for a collection of very readable introductory essays on ecological themes.)
Ehrlich, P. (1988) *The Machinery of Nature*. London, Paladin Grafton Books.
(Agreeable introduction to evolutionary and population ecology for the non-specialist by a well-known practitioner.)
Krebs, C. (1988) *The Message of Ecology*. New York, Harper and Row.
(Good introductory treatment of a number of important topics in ecology.)
Trudgill, S.T. (1988) *Soil and Vegetation Systems*, 2nd edn. Oxford, Oxford University Press.
(Good account of the behaviour of chemical elements in terrestrial environments emphasizing the systems approach.)

Part 3 *The history of the biosphere*

In this part of the book we consider changes in the biosphere during its long history, emphasizing the biological significance of the events signalled as important. We shall also consider changes in the physical environment which have had a major influence on the course of evolution. It is hoped that this section will not appear as a mere catalogue of events, but rather that it will leave the reader with an overview of major evolutionary developments in their historical context. Special attention will be paid to that long interval of time – over 3 billion years (1.0 billion = 1000 million) – before organisms were fossilized in abundance. For most of this time the biosphere was dominated by unicellular organisms, but nevertheless it was during this interval that most of the really momentous evolutionary events occurred.

There is currently a great deal of research activity into historical aspects of the biosphere and a variety of disciplines, including the various branches of the geological sciences, atmospheric science and biology, are involved. It is a very exciting field, and one that is fast changing as new evidence, new discoveries and fresh interpretations challenge the accepted wisdom and give rise to new theories.

In dealing with the Earth's history, it is extremely important to consider the time-scales involved. In Chapters 11 and 12 we shall be measuring time in billions of years, but towards the end of Chapter 14 we shall be discussing events that occurred a matter of several thousands of years ago, which is just yesterday in geological terms.

The 4.6 billion years that have elapsed since the formation of the solid Earth are divided into three aeons (see Fig. 11.1 on page 135). The earliest of these is the Archaean; it is followed by the Proterozoic about 2.5 billion years before present (bybp), and the Proterozoic is followed by the Phanerozoic about 0.57 bybp. In the first two chapters of this section we consider the Archaean and Proterozoic respectively. In the chapter that follows, the changing physical environment is discussed and the subject of fossilization is introduced. Finally, in Chapter 14, we briefly review some major evolutionary events of the Phanerozoic aeon.

11 *Life in the Archaean*

Organisms have been present on the Earth for much of its 4.6 billion year history. It is not known when the first living cells appeared, but it was almost certainly over 3.5 billion years ago and probably in excess of 3.8 billion years ago. In making this statement we raise numerous questions about the early biosphere. In what kind of environment did life originate? How did the earliest organisms evolve? How did they function and how did they affect their environment? In this chapter we discuss the origins of life and the momentous evolutionary events which mark the first 2 billion years of the biosphere.

Even a quite superficial appreciation of the issues involved requires a grasp of the basic material and terminology that were introduced in the first few chapters of the book, so these might usefully be reviewed.

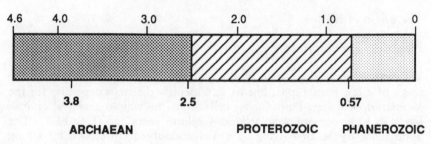

Fig. 11.1 The aeons into which the Earth's history is divided. Life may have been present for almost four billion years; if so, for three-quarters of this time living organisms were primarily unicellular, and for around seven-eighths of this time were confined to water.

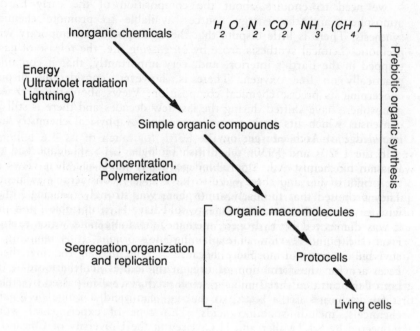

Fig. 11.2   Steps to the origin of life. Considerable uncertainty surrounds the details of
atmospheric composition, the processes involved and even the sequence of some
events.

The origin of life

Prebiotic organic synthesis

We cannot really do justice to this enormously complex subject in the
space of a few paragraphs, but we can identify the steps necessary for the
formation of living, reproducing cells from the simple chemical entities
likely to have been present around 4 billion years ago (Fig. 11.2). The
appearance of the first living cells was undoubtedly preceded by a long
phase of chemical evolution, during which relatively simple molecules of
amino acids and nucleotides were formed, concentrated, and linked, to
produce proteins and nucleic acids respectively. The next series of steps
involved the concentration of these newly-formed polymers, their isolation from the environment and the acquisition of a systematic set of
biochemical functions which constituted a metabolism. The isolation
process implies the development of a kind of membrane, so we are dealing here with the beginnings of cellular formation. Cells also acquired the
capacity to reproduce themselves, although the reproduction of chemical
molecules quite probably preceded cellular formation.

To approach the first problem – the formation of organic molecules – we need to enquire about the composition of the early Earth's atmosphere and the energy sources available to promote chemical synthesis. There is little dispute that the atmosphere contemporary with prebiotic chemical synthesis arose by *outgassing*, i.e. the release of gases trapped in the Earth's interior, and, very importantly, that it contained practically no free oxygen. There is, however, considerable dispute concerning its precise chemical composition. Views about the Archaean atmosphere have shifted during the last few decades and there is still no consensus which fits both an understanding of physical chemistry and knowledge of Archaean geology.

In the 1920s and 1930s the British biologist J.B.S. Haldane and the Russian biochemist A.I. Oparin independently published their views on the origin of life and these were to have a very considerable influence. They proposed that the early atmosphere was strongly reducing (otherwise any simple organic molecules would have been unstable), and that it was dominated by hydrogen, methane, ammonia, and water vapour. From the point of view of explaining life's origins, it is tempting to envisage such an atmosphere because it contains all the ingredients necessary for the formation of organic molecules. Furthermore it has been demonstrated that similar gaseous mixtures, when passed over electrical discharges (as a source of energy) can yield a range of organic chemicals, including amino acids. This type of experimental work, pioneered by S.L. Miller and H.C. Urey at the University of Chicago in the early 1950s, provided evidence, not only for one step in the formation of life but also for the composition of the early Earth's atmosphere.

In the 1950s, American geologist W. Rubey argued that the accepted view of an atmosphere rich in hydrogen, methane and ammonia was not consistent with knowledge of geochemical and atmospheric phenomena, and he proposed a prebiotic atmosphere made up largely of nitrogen, carbon dioxide and water vapour. Under laboratory conditions, such a gaseous mixture proved to be less fruitful in terms of organic products than one containing methane and ammonia, but amino acid synthesis has been achieved. There is evidence in favour of such an atmosphere. First, it is difficult to explain how large quantities of nitrogen could have been added subsequently to the atmosphere. Second, the amount of atmospheric carbon dioxide was probably greater in the Archaean than today; emissions from present-day volcanic events are rich in carbon dioxide, and so too presumably were the emissions from volcanic events which it is believed were much more common during the Archaean. Much of this carbon is now held within carbonate sediments, having come out of the atmosphere in solution. Also, there is evidence that carbonic acid, formed from carbon dioxide and water, was responsible for erosion on the early Earth. Third, methane and ammonia have very short residence times in the atmosphere and so were unlikely to have been present in any quantities.

Ammonia still presents a problem. Although it is a lot easier to envisage the synthesis of organic chemicals in its presence than in its absence, ammonia is extremely unstable in the presence of sunlight. One possibility is that small amounts of ammonia were present locally, perhaps in gas streams emerging from underwater hot springs, in sufficient quantities to permit the synthesis of simple organic molecules. Another suggestion is that ammonia could have been produced from nitrogen and water vapour in the presence of sunlight when titanium was available as a catalyst. Such a reaction can occur in very hot deserts where titanium is present in sand grains.

A point from this discussion is that in attempting to explain life's origins, it is important to use the best evidence provided by geologists and atmospheric scientists and not be tempted to conclude that the early atmosphere must have approximated that in which organic synthesis occurs most readily.

The synthesis of organic substances from inorganic chemical entities requires a source of energy of course, first to promote the dissociation (splitting-up) of existing gaseous molecules, and second to stimulate reactions leading to the synthesis of new molecules. The two most likely energy sources on the early Earth were lightning discharges and ultraviolet radiation. The intensity of ultraviolet radiation would have been much greater four billion years ago than now because there was no ozone in the atmosphere, ozone formation being dependent on free oxygen. Despite disagreements about the composition of the Archaean atmosphere and the energy sources responsible for the synthesis of organic molecules, it is probable that the initial processes occurred in the atmosphere and that the products were deposited in the oceans. The next step to consider is how these simple organic chemicals were linked together to form polymers such as proteins and nucleic acids.

The initial formation of organic polymers required a mechanism for the concentration of the chemical building blocks, notably amino acids and nucleotide units. In the open ocean such molecules would not have existed in a sufficiently concentrated form for their chemical linkage to take place. Concentration might have occurred in shallow water where evaporative losses greatly exceeded inputs or, possibly, by freezing of water bodies, with consequent coalescence of organic materials. Alternatively, polymerization could have occurred as a consequence of organic chemicals being removed from solution and adsorbed on the surfaces of clay particles by electrochemical forces. The selective adsorption of some nucleotides and amino acids has been demonstrated experimentally, so it is plausible that such a mechanism was involved in the synthesis of key biological molecules.

So in summary, prebiotic chemical synthesis is envisaged as a three-stage process involving, first, the production of quite simple organic chemicals, second, their concentration and, third, their polymerization. It is very unlikely that such a series of processes could occur today. Ultraviolet radiation, a possible energy source for the necessary reactions, is weaker due to the screening effects of ozone; any organic chemicals produced abiotically would be unstable in the present oxidizing environment; and anyway our planet now teems with living organisms, even in the most extreme environments, so any newly synthesized organic substance would be quickly consumed and metabolized.

The formation of cells

Living cells are not simply randomly-assembled packages of organic chemicals, but reveal a very high degree of organization. The evolution of even the simplest living cell must have involved a lengthy series of processes and the formation of intermediate stages that we can conveniently refer to as *protocells*. Important steps in the formation of living cells include segregation of polymers from their environment, their organization and interaction in a primitive metabolism and their reproduction. It is not known whether nucleic acids, with their capacity to order biochemical functions through enzyme production, were involved in the earliest energy transformations. Neither is it known how energy, released in the breakdown of organic molecules, was captured and harnessed for use by the earliest cells, or how and at what stage the prebiotic chemical machinery became surrounded by an insulating membrane. It is probable, however, that ATP, 'life's universal energy currency', appeared early, together with the pathway of energy metabolism called glycolysis which we introduced in Chapter 5.

Despite enormous complexities, experimentation has hinted at some possible explanations for a few of the necessary steps. It has been shown that very short peptides (chains of amino acids), adsorbed on to a metal-rich clay surface, can actually catalyze a self-replication process that results in the formation of longer peptides. The initial peptide is thus functioning as a simple enzyme. However, solutions to other problems remain highly elusive. How, for example, could sequences of nucleotides begin to direct the synthesis of proteins from amino acids, a process that is central to the functioning of living systems?

Evidence for the age of the biosphere

Two major lines of evidence are used to address the problem of when life originated on the planet. One involves fossilized remains, the other depends on geochemical information. As recently as 1950 there was very little fossil evidence for life much before about 600 million years ago. Now the fossil record probably extends back about 3.5 billion years. We say 'probably' because of the considerable difficulties in interpreting remains of such great age. Perhaps the oldest fossil cells come from rocks located in the vicinity of North Pole in Western Australia. These show some resemblance to cyanobacteria in both their size and in the way that the cells form filaments. Ancient microfossils, believed to be 3.2 billion years old, also come from the Fig Tree Formation in South Africa. Before these findings, the oldest known fossils were those found on the northern shore of Lake Superior in Canada. The bacteria-like fossils found here were dated at about 2.2 bybp. Their discovery in the late 1950s was very significant, for in the space of a decade, the fossil record had been pushed back some 1.6 billion years.

Geological structures that provide important evidence for ancient life are *stromatolites*. These are roughly columnar or domed-shaped limestone structures that in vertical section reveal alternating bands of organic-rich and organic-poor material. Some stromatolites date to Archaean times although they became much more abundant in the Proterozoic. Localized congregations of 'active' stromatolites are found today in some shallow saline waters in the tropics and sub-tropics, notably off the Western Australian coast and the coast of Baha in northern Mexico. The significance of stromatolites is that they are associated with biological activity. They can form as cyanobacteria colonize surfaces in shallow water; these then trap sediment, but subsequently grow through the sediment to form another organic layer. Some of the earliest fossilized cells are associated with ancient stromatolites, but even the presence of stromatolites is suggestive of biological activity. Circumstantial evidence for the very earliest organisms on the planet comes from some of the oldest sedimentary rocks, dated to about 3.8 bybp. These rocks, of the Isua Formation in western Greenland, contain graphite, a pure form of carbon that is probably of biogenic origin. Further indirect evidence comes from very ancient sedimentary rocks in which the ratio of carbon isotopes is characteristic of living organisms.

The earliest organisms

The conventional view has been that the earliest organisms on the planet were heterotrophic. Such a view accords with a picture of life's origins in water with preformed organic molecules serving as an energy substrate. A recent alternative suggestion is that the first organisms were autotrophic, using a chemosynthetic mode of nutrition (see page 81). This idea is associated with the discovery of chemosynthetic bacteria around hydrothermal vents on parts of the ocean floor. Here, such organisms meet their energy needs by oxidizing hydrogen sulphide, released in solution, in very hot water. Maybe the earliest organisms used a similar mode of nutrition. Hydrothermal vents are possible sites for prebiotic organic synthesis as well, because in such environments a range of inorganic substances is found, and at a very high temperature.

The origin of photosynthesis

If the first organisms were heterotrophic, further evolutionary developments would have been constrained by the amounts of preformed organic substrates that served as an energy source. If, on the other hand, the first organisms were chemosynthetic, further developments would have been constrained by the distribution and amounts of suitable inorganic energy sources. Therefore the appearance of a metabolism that utilized a truly renewable source of energy was crucial. We thus come to one of the most important events in the history of the biosphere, the appearance of photosynthesis.

Recall that in photosynthesis carbon dioxide is reduced by hydrogen, donated either by hydrogen sulphide in the case of certain anaerobic bacteria, or by water in the case of the cyanobacteria and virtually all other photosynthetic organisms. Sulphur is released when hydrogen sulphide is used but oxygen is released when water is the hydrogen donor. The first photosynthetic organisms, which were anaerobic, probably used hydrogen sulphide as a source of hydrogen atoms, like present-day green and purple sulphur bacteria. So for the first time solar energy was being transformed to chemical energy in the biosphere.

The development of a photosynthetic pathway that utilized water as a source of hydrogen, and released oxygen as a by-product of these reactions, was equally profound. Now there was an autotrophic habit which utilized a universally available source of hydrogen. The oxygen generated was to have a determining effect on the course of evolution. Photosynthesis accounts for virtually all the free oxygen in the atmosphere, and from this oxygen ozone is produced. The implication of this last point will become clear later.

Stromatolites are important for recording the development of photosynthesis, as well as providing evidence for the antiquity of life in general, because their formation is closely associated with photosynthetic organisms. As pointed out earlier, some stromatolites originate from earlier than 2.5 bybp, but they do not become abundant until Proterozoic times. Further evidence for the antiquity of photosynthesis comes from comparisons of the ratios of carbon-12 and carbon-13 between fossil remains and the surrounding rocks that are dated to over 3.0 bybp.

Nitrogen fixation

Another highly significant event of the Archaean was the appearance of biological nitrogen fixation. As mentioned earlier, the atmosphere contemporary with the earliest organisms was probably rich in nitrogen gas, and so were the oceans to which life was restricted. In this form, nitrogen is unavailable to most organisms. Only those organisms that possess a particular enzyme system, nitrogenase, can combine nitrogen with hydrogen (see page 124) so that the nitrogen becomes biologically available. Nitrogenase is still confined to prokaryotic organisms and it cannot function in the presence of free oxygen (testimony to its evolution under anaerobic conditions). When oxygen began to accumulate, nitrogen-fixing organisms evolved to metabolize aerobically. However, they also evolved mechanisms to exclude oxygen from the sites where nitrogenase 'fixes' gaseous nitrogen.

12 *Life in the Proterozoic*

The boundary between the Archaean and Proterozoic aeons, set at 2.5 bybp, does not mark a significant change in the record of life. Throughout the first billion years of the Proterozoic, the only life forms present were unicellular prokaryotes, although no doubt these bacteria-like organisms increased their range of metabolic adaptations. By around two billion years ago it seems that free oxygen had begun to accumulate in the atmosphere and then, commencing around 1.5 bybp, certain momentous evolutionary events occurred which we discuss in this chapter.

Photosynthesis and oxygen accumulation

Due to photosynthetic activity, the Earth's atmosphere seemingly became oxidizing, rather than reducing, around 2 billion years ago. Oxygen-releasing photosynthesis had appeared, however, over a billion years earlier. It took such a long time for free oxygen to accumulate because of the reduced nature of rocks and minerals at the surface of the Archaean Earth. Oxygen released in photosynthesis would initially have combined with these substances and accumulate only after these oxygen 'sinks' were full. Furthermore, the oxygen generating capacity of the Archaean biosphere, confined as it was to water and therefore probably nutrient-limited, was only a fraction of today's biota. At current rates of photosynthesis, the present level of atmospheric oxygen could be replenished in a matter of a few hundred years.

Mineralogical evidence has been used to support the view that the Archaean and early Proterozoic atmospheres were essentially reducing and for suggesting the timing of oxygen accumulation. Two minerals, uraninite (UO_2) and pyrite (FeS_2), are readily oxidized further on exposure to the present atmosphere; therefore they should not be present in situations where oxygen is freely available. Where these minerals are found in rocks less than 2 billion years old it seems oxygen was locally absent, but in rocks older than 2 billion years it seems they formed in situations likely to have been exposed to the prevailing atmosphere but were covered and have since remained buried. In contrast, other minerals are so oxidized that they could have formed only in an oxygen-rich environment. Such a mineral is haematite (Fe_2O_3), which has a cementing effect in a variety of rocks. Haematite gives a characteristic red colour to rocks, producing what are referred to as 'red beds'. The age

Fig. 12.1 Banded iron formation. The iron-rich darker bands, 2–3 cm thick, are red in colour, suggesting that the iron was oxidized at the time of formation. Such formations are confined to the Archaean and, particularly, the early Proterozoic.

of the oldest haematite-containing rocks should therefore approximate the beginnings of an oxidizing environment. Haematite is confined to rocks less than about 2.3 billion years old, thus corroborating the evidence from uraninite and pyrite.

Weakly oxidized iron occurs in finely laminated, usually siliceous, structures known as *banded iron formations* (Fig. 12.1). These formations are characterized by bands containing oxidized and reduced iron. Although the processes involved in their formation are complex, the important point here is that some of the iron present was seemingly oxidized when it was originally incorporated. Banded iron formations are unique to Archaean and early Proterozoic times so provide good evidence for early photosynthesis and the gradual accumulation of free oxygen

Initially, molecular oxygen posed a threat to anaerobic organisms; to them the 'new' gas was poisonous. Only those organisms that excluded oxygen from the sites of metabolism, or lived in anoxic situations, could survive. Later, organisms developed a metabolic pathway that utilized oxygen which, as discussed in Chapter 5, are a lot more efficient than anaerobic energy metabolism. Free molecular oxygen was therefore a very potent selective force in early Proterozoic times.

The appearance of the eukaryotic cell

The most momentous evolutionary event of the Proterozoic was unquestionably the appearance of the eukaryotic cell. We have already outlined (page 63) the major points of distinction between prokaryotic and eukaryotic organisms and stressed how fundamentally different they are. To recap, eukaryotic cells possess paired chromosomes located within a membrane-bound nucleus; they have a variety of other subcellular structures with specialized biochemical functions, and a capacity for proper sexual reproduction, associated with meiosis, which halves the number of chromosomes in the nucleus (page 32).

Evidence for the timing of the appearance of eukaryotic cells is based on the comparative sizes of modern prokaryotic and eukaryotic cells and the structural details of fossil cells. Fossil cells older than about 1.5 billion years are always much smaller than is typical for eukaryotic cells. However, sediments from about this age have yielded much larger cells suggesting the emergence of eukaryotic organisms. Dimensional clues are supplemented by evidence of organelles within fossilized cells around 1.5 bybp, although of course there are difficulties in interpreting the details of cellular structure in remains of such great age.

It is not known whether environmental conditions prevented the evolution of the eukaryotic cell before about 1.5 bybp. Because eukaryotic organisms require oxygen, it could be that their origins coincided with oxygen levels reaching a critical minimum threshhold. Possibly this evolutionary step was not possible previously for other environmental reasons, or perhaps it was the pace of evolution, independently of environmental conditions, which was the most important factor.

The question as to *how* eukaryotic organisms arose from their prokaryotic predecessors is not easily answered. A little before the turn of the present century, after advances in light microscopy enabled closer inspection of subcellular structure than previously, the resemblance of the cellular nucleus to some bacteria was noted. Some biologists conjectured that the nucleus had a bacterial origin. This radical proposal received a little support but was never accepted very widely. From the mid-1970s, however, the idea that organelles, and not only the nucleus, owe their origin to other organisms has been resurrected. It has been propounded with considerable enthusiasm by the American biologist Lynn Margulis and now there is widespread support for this theory.

Essentially, this 'symbiotic' theory of the origin of the eukaryotic cell holds that prokaryotic organisms devoured, but failed to digest, other prokaryotic organisms, and the latter ultimately evolved into the organelles such as mitochondria and chloroplasts. In support of this view is the fact that some DNA is usually present in mitochondria and chloroplasts as well as in the nucleus. Furthermore a very 'simple' eukaryote, a species called *Pelomyxa palustris*, has no mitochondria; its respiratory functions appear to be carried out by symbiotic bacteria which live within its single cell. Not surprisingly, there are still many conceptual difficulties with this theory: there is no really satisfactory explanation for the origin of the nucleus with its multiple, paired chromosomes. But whatever the processes involved in its origins, the appearance of the eukaryotic cell had enormous implications for the biosphere, heralding the advent of sex and the appearance of multicellular organisms.

The evolution of sex

As we discussed in Chapter 3, prokaryotic organisms reproduce asexually: a copy is made of the DNA molecule that comprises the single chromosome and later the cell divides to form two new cells. Each of the daughter cells is therefore genetically identical to the mother cell. Some limited exchange of DNA may occur between prokaryotic cells, but the pace of evolution is constrained by the rate of changes in the DNA, that is, the mutation rate. Many eukaryotic organisms also reproduce asexually, and again the offspring should be genetically identical to the parent. It is assumed that the earliest eukaryotes reproduced only asexually, but at some later stage, maybe over a billion years ago, sexual reproduction appeared.

Sexual reproduction requires meiotic nuclear divisions, during which chromosomes sort independently and genes may be exchanged during crossing-over (see page 34). These processes serve to promote genetic variation. When two parents are involved, half the genetic material of the zygote, i.e. the first cell of the new generation, is provided by one parent and half by the other parent. Consequently, new permutations of genes are continually being thrown up on which natural selection can operate. Sexual reproduction therefore offers the possibility of a faster rate of evolution than asexual reproduction alone, which should be particularly advantageous in a changing environment. To reinforce an important point made in Chapter 3, the significance of the appearance of sexual reproduction lay not in reproduction itself, but in the promotion of genetic variability.

The emergence of multicellular organisms

All the types of organism we have referred to so far in this part of the book are essentially unicellular. Unicellular organisms, even though they may form colonies or filaments, operate as independent entities whereas a multicellular organism reveals a considerable degree of functional interaction and interdependence between its cells, tissues and organs. It is believed that the three, almost exclusively, multicellular kingdoms, the plants, the animals and the fungi, arose independently from unicellular eukaryotes in the later stages of the Proterozoic, and more than once in each kingdom.

There are several advantages to multicellularity. First, a multicellular body provides considerable scope for diversification in design. Second, multicellularity enables enormous increases in size of an organism with increased scope for regulating its internal environment. Third, a multicellular body permits cellular specialization and the formation of specialized tissues and organs with particular functions. Fourth, as new cells are produced to replace ageing cells, a multicellular organism can live far longer than a unicellular organism.

Two basic models explain the origin of multicellular organisms from

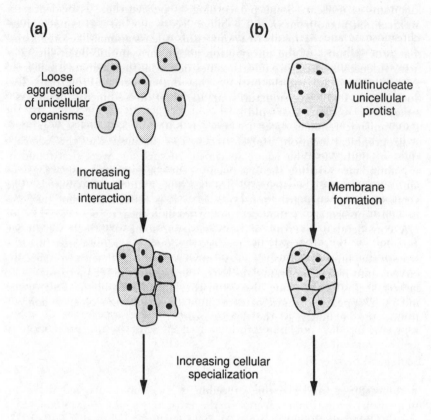

Fig. 12.2 Two models for the origin of multicellular organisms. One (a) envisages closer
intercellular coordination and interdependence; the other (b) envisages
membrane formation within a multinucleate cell.

their unicellular ancestors (Fig. 12.2). One envisages that a number of
separate unicellular organisms became increasingly closely associated,
leading to a greater degree of mutual interdependence until the capacity
to exist independently was finally lost. Alternatively, nuclear division
might have occurred within a single cell, followed by the formation of
membranes to separate nucleated 'compartments'. Currently, the former
of these scenarios is favoured, largely because some of the simplest
animals, such as sponges, behave in many ways like a loosely co-
ordinated colony of cells. Evidence for other types of animal is less
persuasive though, and for them a multinuclear origin may be more
likely.

The fossil record does not permit the origins of multicellular life to be dated accurately, and its early development in the three multicellular kingdoms is unclear. Some palaentologists believe there is evidence for multicellular algae from over 1.3 billion years ago but this is not universally agreed. It is the fossils of animals, or at least animal-like organisms, that provide most of the information about early multicellular life. The first evidence for these organisms comes not from their remains, but from the tracks they laid and the burrows they dug in the late Proterozoic seafloor. Such fossilized signatures appear first in sediments about 900 million (0.9 billion) years old.

Organismic remains dated to between 680 and 580 mybp, regardless of geographical location, are referred to as Ediacaran after the Ediacara Hills in South Australia where specimens of such age were first found in any abundance. During the last twenty years or so, fossil assemblages of similar age have been found to be quite widely distributed. The organisms were primarily soft-bodied, so it is remarkable that they are as well represented as they are in the fossil record.

A considerable diversity of body designs, some of which were once thought to be ancestral to modern groups, is represented in late Proterozoic faunas. However, it is now recognized that most failed to survive into the Cambrian, the first period of the Phanerozoic. The earliest shelly fossils come also from the later stages of the Proterozoic and their appearance anticipates a rapid diversification of such animals around the beginning of the Phanerozoic.

Ecological perspective

In this chapter, and the one preceding it, we have surveyed the most momentous evolutionary events of the biosphere's first 3.2 billion years or so. The approximate timings of these events are shown in Fig. 12.3. It is worth re-emphasizing that during this enormously long interval, which accounts for around eighty-five per cent of the biosphere's entire history, the only organisms present were unicellular prokaryotes. Watersplitting photosynthesis evolved quite early in the history of life, and it seems that as a result free oxygen was accumulating in the atmosphere by 2 billion years ago. By the end of the Proterozoic, however, the atmospheric oxygen concentration was probably no more than twenty per cent of its present value. The accumulation of oxygen enabled the development of an oxygen-using energy metabolism which is far more efficient than fermentation. Throughout the Archaean and Proterozoic, all organisms were still confined to water.

Until the appearance of eukaryotic organisms, ecological communities comprised only primary producers and decomposers. The appearance of eukaryotic organisms, specifically those which made their living by grazing on the photosynthetic cyanobacteria, introduced a new functional dimension into these comparatively simple communities. Using an ecological principle that moderate levels of grazing tend to increase

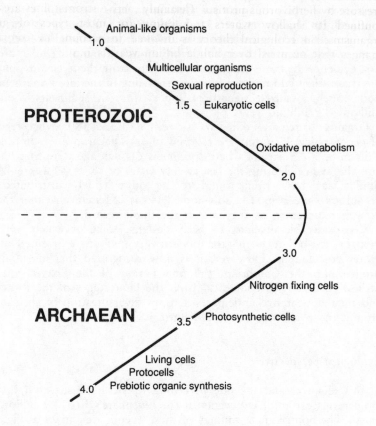

Fig. 12.3 Suggested approximate timings of 'first appearances' in the biosphere during the Archaean and Proterozoic aeons. Numbers are billions of years before present.

species richness, the American palaeontologist Steven Stanley has suggested that unicellular eukaryotes, grazing on relatively uniform beds of cyanobacteria, encouraged the establishment and survival of novel forms by suppressing the growth of the most abundant forms. Thus, diversification would have been enhanced within the primary producer level, leading to increased evolutionary opportunities among the grazing protists. This process may thus partly explain the marked increase in biological diversity which occurred during a relatively short interval commencing around 570 million years ago.

This theory is supported by the history and present distribution of stromatolites, those pillar-like formations referred to earlier.

Stromatolites underwent a worldwide reduction in range during the late Proterozoic, which might have been due to an increase in grazing pressure by herbivorous protists. Certainly, active stromatolites are today confined to shallow waters too saline for most types of grazing organisms. So ecological theory is invoked to account for community changes that occurred over half a billion years ago.

13 *The Phanerozoic: environmental background and fossilization*

This chapter, and the one that follows, concentrate on the most recent 570 million years of biosphere history, the interval known as the Phanerozoic aeon. In the present chapter we introduce the divisions of the Phanerozoic, continental drift and climatic change, fossilization processes and the dating of fossils. Some of these themes are of course relevant to earlier ages as well but they are particularly applicable to evolutionary events in the Phanerozoic, which is the focus of the final chapter.

Divisions of the Phanerozoic

The term Phanerozoic, meaning 'abundant life', is used because sedimentary rocks deposited in this interval of time are very much richer in fossils than those deposited earlier. The lower boundary of the Phanerozoic marks the appearance in abundance of organisms with hard body parts.

The Phanerozoic comprises three *eras*, each of which is further divided into a number of *periods* (Fig. 13.1). Notice that the Palaeozoic, the earliest era of the Phanerozoic, is five times as long as the Cenozoic, the most recent. The actual dates of the boundaries between named periods and eras will be seen to vary between schemes, but not to any appreciable extent. For those unfamiliar with the divisions of the geological time-scale it is probably a good idea to first learn the duration and boundaries of the three eras, and use this framework to examine the periods within each. A little time spent on learning the timing and the terminology will be well rewarded.

The temporal divisions of the Phanerozoic are based on the history of life as revealed by the fossil record. Boundaries between successive eras and periods are marked by significant changes in fossil assemblages. This system, in which most names represent the geographical locations in which the geological formations were first described, developed during the nineteenth century in western Europe. At this time there was no way of dating the rocks or fossils; absolute dating became possible only with

Era	Period	
CENOZOIC	Neogene	
		25
	Palaeogene	
65		65
MESOZOIC	Cretaceous	
		145
	Jurassic	
		210
	Triassic	
245		245
PALAEOZOIC	Permian	
		285
	Carboniferous	
		370
	Devonian	
		410
	Silurian	
		440
	Ordovician	
		505
	Cambrian	
570		570

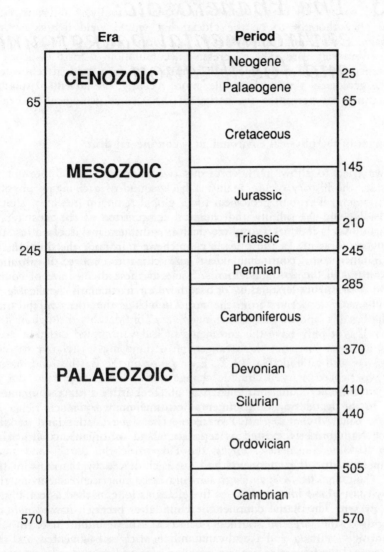

Fig. 13.1 The eras and periods of the Phanerozoic aeon. Numbers are millions of years before present. Location of boundaries varies slightly between schemes.

an understanding of radioisotopes in the present century. However, the *ordering* of events could be inferred from the relative positions of rock strata and the types of fossils they contained. For example, rocks of the Cambrian period were first described in detail in North Wales in the British Isles (Cambria is the Roman name for Wales). Thereafter, rock strata containing similar fossil assemblages, whatever their location, were deemed to be part of the Cambrian system. Actually, Cambrian rocks were once thought to be the oldest that would yield fossils as they contained only comparatively simple life forms and fossil finds in older rocks were very rare. The progressive accumulation of fossil evidence in the nineteenth and early twentieth centuries led to a gradual refinement of the geological time-scale, while, more recently, the advent of suitable techniques has permitted the dating of the various geological intervals.

Changes in the physical environment – continental drift

As we tried to show in the previous two chapters, it is important to consider the history of life against a background of a changing physical environment. Through geological time, global temperatures have altered markedly, and the salinity and chemical composition of the oceans have changed. Such changes in the physical environment inevitably affect the organisms present; perhaps leading to changes in their distribution, or precipitating extinctions, but always influencing the course of evolution by shifting the nature and extent of selective forces.

Much of our understanding of long-term environmental change during the Phanerozoic derives from the sure knowledge that the configuration of the Earth's land masses has changed very considerably throughout this interval. Not only have the continents of today occupied different positions in times past, but the arrangement of land masses relative to each other has also altered (Fig. 13.2). Each point on the Earth's land masses has experienced marked climatic change during the Phanerozoic, due in part to climatic changes at each part of the Earth's surface, but partly also to the wandering of continents across climatic zones.

It is believed that by late Proterozoic times the Earth's land masses, which had formerly existed as separate island continents, coalesced to form a 'supercontinent'. During the Palaeozoic this large land mass fragmented, but the continents had again coalesced by the close of the era. Then, in Mesozoic times, fragmentation again occurred, giving rise to two large land masses. One of these, *Laurasia*, consisted essentially of the present northern continents while the other, *Gondwanaland*, comprised the present southern continents plus India. Late in the Mesozoic, Laurasia and Gondwanaland themselves fragmented and the continental configuration we are familiar with today gradually emerged. However, continental movement has continued throughout the Cenozoic, and indeed is still continuing today.

Continental movement had been seriously proposed by the German scientist Alfred Wegener near the beginning of the present century, but it was not until the 1960s that the phenomenon became almost universally accepted. Central to the modern theory of continental drift is the fact that the *lithosphere*, the rigid outer part of the Earth, is composed of separate segments, called plates. These plates, which comprise both crust and outer mantle material, 'float' on the denser, but more fluid, material of the *asthenosphere* beneath (Fig. 13.3). Note that the thickness of

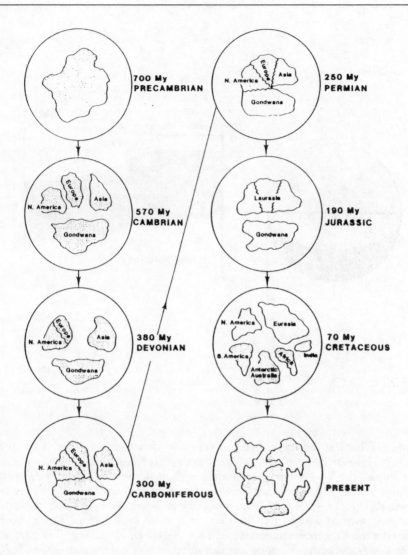

Fig. 13.2 Schematic depiction of the Earth's changing geography during the Phanerozoic.

the plates is greater under the continents than under the sea-floor. There are eight major plates and several minor ones (Fig. 13.4), but the number and configuration of plates has changed during geological time. Notice that not all plates support continents.

It is movement of the plates relative to each other that is responsible for continental drift. Plate movement is caused by the extrusion of molten material along narrow, linear zones on the sea-floor, and by the sinking (*subduction*) of plate material elsewhere. Zones of extrusion

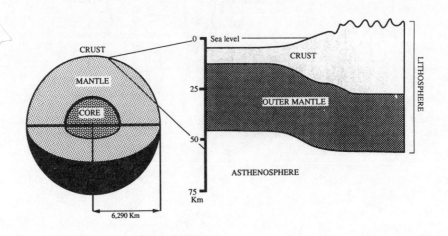

Fig. 13.3 The layered structure of the Earth. The rigid lithosphere is made up of crust
and the thin outer layer of the mantle.

and subduction represent different types of *plate margins* (Figs. 13.4 and
13.5), known respectively as *constructive* and *destructive* margins. For
example, the long submarine mountain chain known as the Mid-Atlantic
Ridge lies at the boundary between the North American and Eurasian
plates (Fig. 13.4). From here, material is being extruded and the sea-
floor is moving away from the ridge at a more or less equal rate in both
directions, a phenomenon referred to as *sea-floor spreading*. The rate of
sea-floor spreading is quite measurable; it has only taken around 70
million years for the Atlantic to open up to its present width.

As new material is extruded on to the sea-floor, it cools and thickens,
and when its density exceeds that of the mantle beneath, it sinks (Fig.
13.5). Earthquakes are an expression of such sinking and are associated
particularly with subduction zones. The continual extrusion and removal
of oceanic crust means that no areas of the sea-floor are much older than
about 200 million years. Parts of all continental land masses are very
much older than the rocks of the sea-floor, the reason being that these
rocks are not sufficiently dense to sink back into the mantle.

At a third type of plate margin, two plates scrape past each other
without extrusion or removal of material. Such a movement produces a

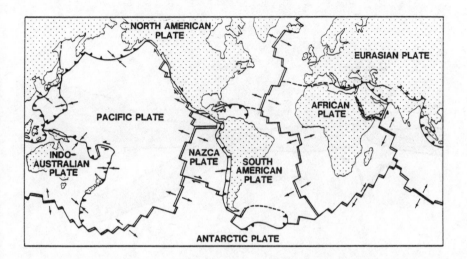

Fig. 13.4   The major plates of the Earth and types of boundary. Constructive boundary
(◿◿); destructive boundary (▲ ▲ ▲); transform fault (⇆); inactive or
poorly defined boundary (- - - -); arrows indicate principal direction of plate
movement.

transform fault; an example is the well-known San Andreas fault which
runs through California.

Continental movement and climatic change

It is easy to appreciate how a latitudinal shift in the position of a continent will have direct climatic consequences for each part of the land
mass. But changes in the size of land masses as a result of their cleavage
or coalescence will also affect rainfall and temperature patterns as the
maritime influence alters. Plate movement can influence climate indirectly
where mountain building is a consequence of plate collision: particular
areas may become increasingly arid as mountains block the movements
of moisture-laden air.

Very importantly, the configuration of land masses affects climate by
influencing the directional movement of ocean currents which transfer
heat energy from the tropics to higher latitudes. The present configuration of the continents permits the circumpolar movement of water

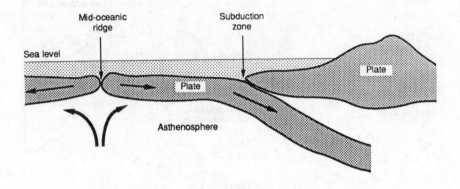

Fig. 13.5 Types of plate margin. In this example, new crust is formed at the mid-oceanic ridge as molten material is extruded and cools. At the subduction zone, one plate is being forced under the other. The direction of plate movement is indicated. The continental plate remains on top due to the relatively low density (greater buoyancy) of continental material. The broad arrows represent the convective forces that are believed to be responsible for the extrusion of material and consequent sea-floor spreading.

between Antarctica and the other southern continents, while in the northern hemisphere the Arctic Ocean is virtually surrounded by land. Thus, both polar regions are more or less isolated from the warming influence of currents moving from low to high latitudes. At times in the past, after land masses had coalesced to form one 'supercontinent' (Fig. 13.2), the circulation pattern of surface currents was probably much simpler than today; equatorial currents could move relatively unimpeded into high latitudes and so the temperature gradient between equator and poles would have been less marked.

Astronomical events such as slight changes in the Earth's orbit and the tilt of its axis may have triggered changes in climate, although there is considerable uncertainty concerning the magnitude of such influences. Whatever the contribution of these phenomena, there is no question that the climate at every location on the planet has changed markedly, and that such changes have had a profound effect on landscape processes and the nature of ecological communities.

As we have implied earlier, the only direct record of life in the past comes from the fossils, so before looking at evolutionary events in the Phanerozoic we briefly consider fossilization processes

Fossils

The term *fossil*, applied originally to anything dug up out of the ground, is now confined to organismic remains or some other tangible evidence of an organism's existence. Remains may be preserved unaltered, but more usually they are preserved in chemically-altered form. Organisms are dying all the time of course, which raises the question of how old a dead organism has to be before it can be considered a fossil. The fact is, we cannot define a fossil very satisfactorily in these terms, but failure to do so is not an issue. The term *subfossil* is sometimes used to refer to remains younger than a few thousand years.

Sources of bias in the fossil record

It is extremely important to appreciate that fossils provide an incomplete, and highly biased, record of life on Earth. First, the fossil record is heavily biased towards organisms with hard body parts such as bones, shells and teeth. Not only are these parts, because of their mineral component, relatively resistant to decay, but they may be further mineralized as chemical substances replace the original organic structure. Chemical changes involving the recrystallization of existing material may also occur to enhance resistance to decay.

Even hard body parts, unless highly mineralized, will eventually disappear if normal decomposition activities continue. Thus the environment in which an organism died (or to which it was transported after death) is critical. By far the most important environments for fossilization are those in which sediments are rapidly accumulating, thus burying organic remains. Effectively, rapid sedimentation is a prerequisite for fossilization as it reduces both biological decay and physical weathering. Burial can be sudden, as in an ash or lava flow, during a desert sand storm, or while a river is in flood. Where burial occurs gradually, preservation depends on the maintenance of low oxygen levels so that decomposition is inhibited. Fossils are therefore associated principally with *depositional* environments, particularly alluvial lowlands and adjacent shallow waters. The geographical extent of environments conducive to fossilization is therefore a second major source of bias in the fossil record.

Fossilized remains of organisms that inhabited upland environments are quite rare, and fossilization occurs only in limited situations. Some limestone caves have yielded fossil finds of bones and teeth; the animals having entered the caves voluntarily before dying, been carried by predators, or simply fallen down shafts.

The chemical and physical processes which combine to convert soft

Fig. 13.6   Trace fossil. A feeding trace left by the worm *Nereites*; from the Silurian, South Wales.

sediments into hard rock, collectively referred to as *diagenesis*, have an important bearing on the fate of embedded organic remains. Diagenesis introduces a further source of bias into the fossil record because organisms vary in the degree to which their remains survive such processes. Metamorphic processes (i.e. changes in rocks due to heat and pressure) and tectonic events (earthquakes, etc.) tend to reduce the integrity of organic remains and reduce the probability of fossilization.

Time is another source of bias in the fossil record. In general, sediments become increasingly altered with age so that the younger the sediments the greater the probability that remains will survive. As we delve further and further back in time the fossil record therefore becomes less and less informative.

Some fossil types

Trace fossils are preserved impressions, left by moving organisms. The size and shape of trace fossils (Fig. 13.6) such as footprints or burrows provide clues to the morphology and life-style of the organisms that left them. Other fossils are moulds of organisms. A body, or an organ, may disappear some time after it has become buried by sediments leaving a void which reveals its outline features. Alternatively, the inner parts of a shelled organism may disappear rapidly, but the shell survives long enough for its interior space to fill with minerals. Disappearance of the outer shell then leaves a mineral replica of the organism.

The remains of hard body parts may survive virtually intact, or the original mineral structure may be partly or wholly replaced (Fig. 13.7). In some fossils, the entire organic structure has been replaced by minerals

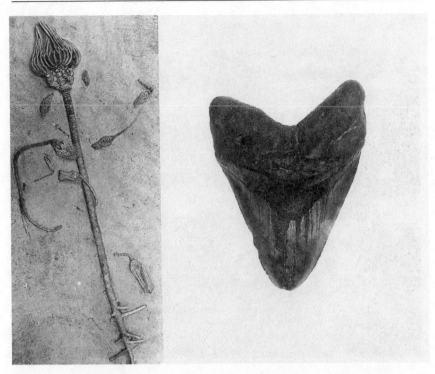

Fig. 13.7 (a) A crinoid (sea lily); from the Carboniferous, Indiana, USA. The outer calcareous material has been preserved but further mineral deposition has occurred within the original structure. This specimen stood 'upright' as a sedentary inhabitant of the sea-floor. (b) Tooth of the shark *Charcarodon*; from the Miocene, Florida, USA. The original mineral structure has been preserved virtually unchanged.

in solution, thus preserving the organism's original form. This process accounts for the survival for well over 200 million years of the famous petrified logs in northern Arizona (Fig. 13.8), south-western United States. Some fossils consist of carbon films: fossilized plant leaves are commonly of this type (Fig. 13.9).

Important fossil finds have been made in some unusual situations. At 'tar pit' locations in California, bones of a variety of mammal types from the comparatively recent past have been found embedded in asphalt. This petroleum-like substance seeped to the surface and trapped dead remains as it hardened. In Alaska and Siberia, animal carcasses have been extracted from the frozen silt of the permafrost zone; preservation here being attributable to rapid freezing and drying on death. One fairly recent find in Siberia to receive much publicity was a young mammoth which died some 40,000 years ago. Preservation was so good that tests and analyses could be carried out on tissues such as muscle, blood and hair.

Fig. 13.8 Petrified logs of coniferous trees of Triassic age, Arizona, USA. Silica has replaced the original carbonaceous structure.

Fig. 13.9 A carbon-film fossil. A fern leaf from Carboniferous coal deposits, north-west England.

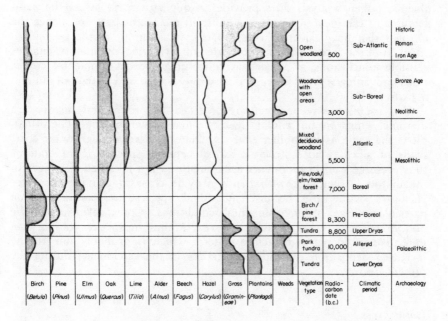

Birch (Betula)	Pine (Pinus)	Elm (Ulmus)	Oak (Quercus)	Lime (Tilia)	Alder (Alnus)	Beech (Fagus)	Hazel (Corylus)	Grass (Gramin- eae)	Plantains (Plantago)	Weeds	Vegetation type	Radio- carbon date (b.c.)	Climatic period	Archaeology
											Open woodland	500	Sub-Atlantic	Historic Roman Iron Age
											Woodland with open areas	3,000	Sub-Boreal	Bronze Age Neolithic
											Mixed deciduous woodland	5,500	Atlantic	Mesolithic
											Pine/oak/ elm/hazel forest	7,000	Boreal	
											Birch/ pine forest	8,300	Pre-Boreal	
											Tundra	8,800	Upper Dryas	
											Park tundra	10,000	Allerød	Palaeolithic
											Tundra		Lower Dryas	

Fig. 13.10 A general pollen diagram for southern Britain during the Holocene. The relative abundance of pollen of different taxa (horizontal axis) is shown through time (vertical axis). Such diagrams are used to reconstruct vegetation history and can be used to infer environmental conditions and human activities.

Fossilized pollen grains, which in general are quite durable, are particularly useful for elucidating the evolutionary history of the seed plants, and also for examining vegetation change during the comparatively recent past. The value of pollen grains is that their morphology varies between plant taxa, while their durability is attributable to a decay-resistant outer coat, the *exine*.

Because pollen analysis is such a valuable and standard technique in palaeoecology it merits further comment here. The usual approach is to extract cores from a suitable depositional environment, particularly a peat deposit or a soft sediment beneath a water column. As these deposits have accumulated over time, increasing depth down a vertical core represents increasing age. The core is dated radiometrically at various levels, and its pollen extracted and examined. In practice it will often be possible to assign a pollen grain only to a genus or family, and not to a species. Also, there are problems in interpreting the pollen record in terms of the species composition of the vegetation. The pollen may have been transported over a range of distances, and furthermore, different plant species produce different quantities of pollen. Nevertheless, pollen analysis does provide an opportunity to assess the main features of the local vegetation at times in the recent past and hence the changes that have occurred (Fig. 13.10). In addition, as the taxa identified from the recent pollen record are still living, pollen analysis enables sensible conclusions to be drawn concerning the environmental conditions prevailing at the time, and also the nature and extent of human activities.

Also useful for reconstructions of recent environmental history are diatoms, which are unicellular algae of aquatic environments. The principle is the same as for pollen analysis. Diatoms have a tough outer coat, made of silica, whose structure varies between species. Dead diatoms thus preserve well in sediments accumulating on a lake bed. Changes in the diatom assemblage with depth provide clues to changes in water chemistry with time because individual species are confined to a particular range of environmental conditions. Cores taken from the bottom of some lakes in Britain reveal a shift in species composition indicative of increasingly acidic waters, which is probably due to the increased deposition of acidic substances in the catchments during the present century.

Dating fossils

As mentioned earlier, during the nineteenth century, rock strata were classified sequentially, but their absolute ages were unknown. Accurate dating has only been possible during the present century with the discovery of radioactivity and an understanding of the behaviour of radioisotopes. The isotopes, including the radioisotopes, of each element occur in characteristic ratios. However, the atoms of radioisotopes are inherently unstable; they disintegrate with the emission of radiation, subatomic particles or both. As atoms of radioisotopes decay, atoms of other isotopes, known as 'daughters', are produced. Rates of decay, which vary enormously between isotopes, are expressed as *half-lives*; the half-life of an isotope being defined as the length of time it takes for half the radioactivity to disappear. The ratio of a parent isotope to a daughter isotope in a substance permits the age of formation of that substance to be calculated and is the basis of radioactive dating.

As half-lives vary so much, a particular 'mother-daughter' system can be used confidently only for a particular time-scale. The system rubidium-87–strontium-87 is used for very long time-scales on account of the extremely slow decay rate of the parent isotope (appoximately 48.5 billion years) and the inherent stability of the daughter. Potassium-40–argon-40 is also employed for dating old rocks because potassium-40 has a half-life of around 1.3 billion years.

For organic remains no older than a few tens of thousands of years, radiocarbon-dating is used. Age is calculated on the basis of the ratio of two isotopes of carbon; one is carbon-14, a radioisotope with a half life of 5730 years, the other is the stable isotope carbon-12. Carbon-14 is naturally present as a very tiny and characteristic proportion of carbon-12, and the method assumes the two isotopes are assimilated in photosynthesis in this same ratio. Carbon-14 will then decay (to produce an isotope of nitrogen), so by measuring the ratio of the two carbon isotopes it is possible to date the age of the carbonaceous material; the lower the amount of carbon-14 relative to carbon-12 the greater its age, because more radioactive carbon will have decayed.

When the time interval occupied by a fossil taxon is well known, it may be used to date rocks. Fossil types in common use for dating purposes are called *index fossils*. The ideal index fossil would be easily and unambiguously identified, be abundant and widely distributed in different rock types, and have lived for a clearly defined and comparatively short time interval. In practice all these requirements are not often met, and although fossils are routinely used for dating rocks, there are problems with this approach. One source of error is that the geographical distribution of a taxon is liable to vary with time, which makes accurate dating difficult.

14 *Life in the Phanerozoic*

In this chapter we review some of the major evolutionary events of the most recent 570 million years. We consider the three eras in turn and within each identify topics, more or less in chronological order, and consider their evolutionary significance. This means some tracking back in time as well as moving forward, but it will permit topics to be developed further when they are first raised. It will be useful to periodically consult the geological column shown in Fig. 13.1 to place events in a temporal context and to check the Earth's historical geography from Fig. 13.2. Before commencing this review we briefly consider the pace of evolution as revealed by the fossil record.

Extinction and adaptive radiation

Despite the fact that the fossil record is so incomplete and biased towards hard-bodied organisms, some generalizations can be made about the history of life. One of these is that the appearance and disappearance of taxa has not proceeded at a uniform rate over the last half billion years or so. Periodically, taxonomic diversity has declined very abruptly in geological terms, while at other times taxonomic diversity has rapidly increased.

The more or less sudden disappearance of a significant proportion of taxa is termed a *mass extinction*, and it is believed that five such events took place during the Phanerozoic. The most dramatic of these occurred in shallow water environments some 245 million years ago and marks the close of the Permian period and the Palaeozoic era. Perhaps the most commonly-known, however, marks the close of the Cretaceous around 65 million years ago, when the dinosaurs disappeared. The other three mass extinctions occurred around 440, 370 and 210 mybp and mark the close of the Ordovician, Devonian and Triassic periods respectively.

But in addition to the five big ones, other extinction events have occurred. D. Raup and J. Sepkoski of the University of Chicago have analysed a massive palaeontological data set and concluded that such events do not occur randomly through time, but with a periodicity of somewhere around 26 million years. There is by no means complete agreement among palaeontologists about this conclusion, nor about the explanation. Regularity of extinction events really calls for a single primary cause, and regular events are best explained in terms of cosmic, i.e. extraterrestrial, phenomena. One suggestion with a degree of support presently is that extinctions followed collisions with asteroids or cometary showers. We return to this intriguing possibility when we consider the mass extinction at the close of the Cretaceous.

The Phanerozoic is also marked by intervals of rapid evolutionary diversification, known as *radiations*. Radiations follow significant extinctions as heavy attrition among an existing biota offers considerable evolutionary opportunities for the survivors. Important radiations therefore follow mass extinctions, but what about the pace of evolution at other times? On this subject there is more debate. Typically, species appear relatively suddenly in the fossil record, remain more or less the same for some 1 to 3 million years and then become extinct. Such observations are not in themselves particularly contentious. But Darwinian evolutionary theory predicts slow, gradual change, explained now in terms of gene-by-gene replacement. New species arise therefore by progressive modification of existing forms. Failure to observe speciation in the fossil record is conventionally attributed to the fact that it is so incomplete.

But if the fossil evidence is taken as a faithful record of the history of life, can present evolutionary theory account fully for the observations concerning new species, which appear quite abruptly, or must an additional mechanism be involved? On this issue there have been fierce debates, sparked largely by Harvard University's Stephen Jay Gould and Nils Eldredge of the American Museum of Natural History in New York. In the early 1970s they proposed the term *punctuated equilibrium* for an evolutionary pattern characterized by long intervals of little morphological change punctuated by brief – i.e. measured in thousands rather than millions of years – intervals of rapid evolution. More contentiously, they also suggested that these episodes of rapid morphological evolution may require more than cumulative single gene replacement, and that as a consequence of divergence of evolutionary lines, new species arise. It must be said that this is a minority view, and is strenuously resisted by many working with evolutionary mechanisms. Now we attempt to review major events in the Phanerozoic as revealed by the fossil record.

The Palaeozoic era

The early Palaeozoic

The lower boundary of the Cambrian, the first period of the Palaeozoic era, marks the onset of an interval of very rapid diversification of animals of shallow marine habitats, a phenomenon often referred to as the Cambrian explosion. The basic body plans of all the present day animal phyla can be traced to a comparatively brief interval in the early Palaeozoic. Other phyla appeared also, but later became extinct.

A particular feature of the early Cambrian was the appearance of a

(a)

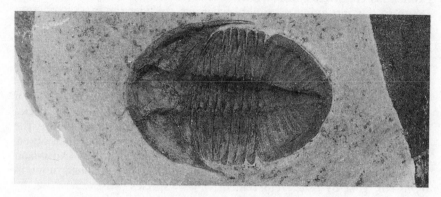

(b)

Fig. 14.1 Two types of invertebrate well represented in Palaeozoic fossil assemblages. (a) An Ordovician trilobite, *Ogyginus*. (b) A Carboniferous brachiopod, *Brachythyrus*.

large variety of organisms with calcified body parts, often external skeletons. These included gastropod and bivalve molluscs (respectively snails and clams), brachiopods, corals and sponges, and the well-known trilobites (Fig. 14.1). The trilobites, early relatives of the insects and crustaceans, are useful for geological dating because the many taxa occupied comparatively short, well-defined time intervals. Towards the end of the Cambrian, the trilobites suffered a wave of extinctions and never regained their former prominence. It is not known for sure what circumstances stimulated the evolution of hard body parts, but two important factors may have been changes in water chemistry, particularly regarding calcium and magnesium concentrations, and more effective predation, against which shells afford some protection. Whatever the factors involved, the biological incorporation of certain minerals (*biomineralization*) on a large scale had important implications, not only in a biological sense, but also for the circulation of certain chemical elements.

Although the record of life in the Palaeozoic is very largely a record of organisms with hard body parts we should not conclude that soft-bodied creatures were unimportant. As discussed in Chapter 13, fossilization processes take time and soft body parts usually disappear quickly. It is only under special circumstances that they are preserved. Rapid burial under almost anaerobic conditions led to the fossilization of a variety of soft-bodied organisms in a tropical marine environment now represented by part of the Rocky Mountains in southern British Columbia, Canada. Here a rock stratum known as the Burgess Shale provides a unique glimpse of soft-bodied life in middle Cambrian times. Within these deposits have been found a large and diverse array of organisms; some seem familiar, but many types are quite alien to our eyes and apparently did not persist for very long. It is rare finds, such as those in the Burgess Shale, that emphasize how incomplete and biased is our picture of living communities in the distant past.

Throughout the Cambrian the principal photosynthetic organisms were eukaryotic, single-celled acritarchs. These first appear in the fossil record from about 1.4 billion years ago but do not become abundant until the later stages of the Proterozoic when they seem to have undergone considerable diversification. The survival of fossilized acritarchs is due to the cysts, or resting stages, that they produced. Towards the end of the Cambrian though the acritarchs rapidly declined. The ostracoderms, primitive jawless fish, appeared in late Cambrian times, first in the oceans but later in fresh-water as well.

The considerable morphological diversification that occurred during the Cambrian must have had significant ecological and evolutionary consequences. In part at least we can justifiably regard the course of evolution as a response to changing relationships among organisms. No doubt ecological communities became more complex as diversity increased and the ecological roles of the various organisms became more specialized. The Ordovician, which follows the Cambrian, was a very important interval for diversification of marine organisms. Its close, however, is marked by one of the five major mass extinction events of the Phanerozoic.

The colonization of land

Today, such extreme conditions are necessary to prevent at least some organisms from colonizing land surfaces that it is quite difficult to envisage landscapes that are devoid of life. Life on land requires sufficient atmospheric ozone to screen out much of the sun's ultraviolet radiation (page 71), particularly the shorter wavelengths. Ozone formation is dependent on free oxygen, which was accumulating, as a result of photosynthetic activity, throughout much of the Proterozoic. With increased absorption of ultaviolet radiation by atmospheric ozone, organisms could live closer and closer to the water surface and eventually on land surfaces. But although ozone formation can be considered an enabling factor in the colonization of land there is no evidence that it was the trigger event.

Land colonization should not be thought of as a single, sudden event, but rather a very gradual process associated with morphological and physiological changes that resulted in some types of organism becoming increasingly less dependent on standing water. After all, the distinction between terrestrial and aquatic organisms is not absolute: some organisms are truly aquatic in the sense that they live their entire lives in water, but many organisms are semi-aquatic, spending part of their life in water and part on land.

For a long time, the mid-Silurian was regarded as a convenient reference for land colonization, but recent discoveries, including trace fossils of centipede-like animals from the Ordovician suggest that some revision of thinking may be necessary. The fossil record is not very revealing about the first photosynthetic land colonists, or their evolutionary history, but they were probably multicellular algae with a semi-aquatic mode of life. Mosses and liverworts (bryophytes), which also require access to copious water, probably evolved from green algae to become early occupants of moist habitats. The fungi too, no doubt, were among the earliest organisms to invade essentially terrestrial environments.

During mid-Palaeozoic times animals also colonized the land, but as emphasized earlier, we should regard land colonization as a very gradual process. The early animal 'land' colonists were semi-aquatic, depending on free water for reproduction. Insects occur in early Devonian fossil assemblages, nearly 400 million years ago. By the late Carboniferous they had developed wings and rapidly diversified to occupy a variety of roles and an ecological status they have never relinquished.

Plant evolution in the Palaeozoic

By the late Silurian, bryophytes had been joined by other, essentially terrestrial, plants. These early land plants had neither leaves nor roots; they were simply branched stems, terminating in enlarged spore-bearing tips. Such plants were anchored in the ground by their stems and were probably confined to marshy areas. Leaves, and most of the key features associated with erect growth on land, appeared during the late Silurian or the succeeding Devonian period. These features include a waxy surface which inhibits moisture loss and thereby prevents desiccation, specialized organs for water and nutrient uptake, and a network of cells (vascular tissue) for the movement of water, assimilates and nutrients. Erect growth can be sustained to a limited extent by water pressure

Fig. 14.2 Reconstruction of an Upper Carboniferous swamp forest. The large plants with jointed stems are sphenopsids, a group now represented by horsetails; the large animal is a labyrinthodont amphibian; the insect is the giant dragonfly *Meganeura*.

within the cellular vacuole and the cell wall which resists cellular swelling, but for plants of any stature a greater degree of structural rigidity is necessary. Increases in plant size were associated with larger amounts of cellulose, and later lignin, being laid down in cell walls. One major consequence of an increasing plant cover, particularly the evolution of true roots, must have been lowered rates of erosion as the land surface became more stable.

The clubmosses and related forms (the lycopods) had appeared by the early Devonian, and may well have been present in the Silurian. The sphenopsids, a group which includes the commonly-named horsetails (Fig. 14.2), appeared in the late Devonian. The ferns probably arose before the close of the Devonian and were certainly present in the Carboniferous. Extant clubmosses are fairly small, but during the Carboniferous some were tree-sized and they contributed to the coal measures that derived from the biomass of late Carboniferous swamps. Carboniferous sphenopsids were also very much larger than the few living representatives of this group, and they appear to have tolerated rather drier conditions than either the ferns or the lycopods.

None of the types of plant mentioned so far produce seeds. The seed plants appeared in the late Devonian. To appreciate the considerable significance of this evolutionary event, it is necessary to review what we learnt earlier (page 36) about the life cycles of plants in general. Recall that there are two distinct generations in the life cycle of a plant, known respectively as the sporophyte and the gametophyte. The sporophyte is diploid whereas the gametophyte is haploid. Which of the two generations is the larger and longer-lived varies between plant types. With mosses and liverworts it is the gametophyte; for clubmosses, ferns, horsetails and all seed plants it is the sporophyte.

The principle of alternation of generations for seedless plants was shown earlier (Fig. 3.5), using a fern as example. A meiotic division of specialized cells on the sporophyte gives rise to haploid spore cells from which the gametophyte, the sexual generation, develops. Some seedless plants produce one type of gametophyte, some produce two, one male and one female, and they originate as different types of spore. It is within the gametophyte generation that the male and female gametes, the sperm and the eggs, are produced. Very importantly, the male gametes of seedless plants require water to reach the female gamete. The key point then is that seedless plants depend on copious water supplies for reproduction.

Now, seed plants produce two types of haploid spore, termed *micro* and *mega*. The smaller microspores, which are male, are later released as pollen grains. The male gametophyte, which carries the male gametes (sperm nuclei), develops from the pollen grain, usually as a tube-like structure. The megaspore, which is female, gives rise to the megagametophyte, in which the egg nucleus is produced. Megaspore development takes place amongst sporophyte tissue which provides protection. A key feature of all seed plants is the retention of the female gametophyte by the sporophyte generation. Compare this situation with that of the fern in Fig. 3.5 in which the two generations are separate entities and the gametophyte is therefore vulnerable, particularly to desiccation.

Pollen that is shed by flowerless seed plants is carried in air. If pollen grains alight in the vicinity of a compatible megagametophyte, the enclosed sperm nuclei are carried towards the egg nucleus. (In the more advanced seed plants the sperm nuclei are carried within a *pollen tube* which extends in response to appropriate stimuli.) A vital point from an evolutionary perspective is that free water is not required for the movement of sperm to egg in seed plants; the only moisture required for this purpose is supplied by the female gametophyte. Successful fusion of sperm and egg nuclei gives rise to a diploid zygote, the first cell of the sporophyte generation. Successive cell divisions then occur to produce a seed.

Seeds are structures that permit development of the sporophyte generation (i.e. the main plant) to be arrested at an early stage. Seeds can be quite durable, most survive well in very dry conditions as they tolerate desiccation, many have features that encourage their dispersal, and under suitable conditions some can remain dormant for considerable periods of

Fig. 14.3   *Ginkgo biloba*, the only extant species of ginkgo, a type of gymnosperm prominent in Mesozoic forests.

time. So seed plants are independent of free water for reproduction, while the seed itself provides a means for preserving an individual until such time as conditions are suitable for its growth and development.

The earliest seed plants were the seed ferns, an extinct group whose frond-like leaves resembled those of the true ferns. Seed ferns date from late Devonian times and were very common in the Carboniferous. One group of seed ferns in particular, the genus *Glossopteris*, was abundant throughout the southern hemisphere continents in late Palaeozoic times and gave its name to a flora which occupied large parts of Gondwanaland in the early Permian. The cordaites, an extinct group of long-leaved, seed-bearing, trees were prominent in late Carboniferous forests.

During the Permian, the conifers (cone-bearing plants) diversified and expanded, largely at the expense of the tree-sized lycopods, sphenopsids and seed ferns. The rise of the conifers is believed to have been associated with climatic changes, to drier or warmer conditions. The ginkgos, a group of seed-bearing trees, originated in the Permian, although seemingly did not become prominent until the Mesozoic. Just one species of this ancient group still survives (Fig. 14.3).

So late Palaeozoic times were very important for the evolution of land plants, including the seed plants. But it should not be thought that seed plants took over completely from their predecessors; the ferns in particular remained a conspicuous group. However we have taken plant evolution a very long way in a short time. Now we return to mid-Palaeozoic times to consider other forms of life.

Animals in the middle and late Palaeozoic era

Fish, or at least fish-like creatures, were the first vertebrate animals. While early Palaeozoic fish were relatively simple, jawless and often armoured, during mid-Palaeozoic times, fish developed more modern features (Fig 14.4). Because of the abundance and diversity of fossil fish from the Devonian, and the evolutionary changes that occurred within the group, this period is frequently referred to as 'the age of fish'. Some types of fish, probably from waters subject to seasonal drying, were the first vertebrates to have lungs, an evolutionary development of enormous significance because it preadapted vertebrates for a terrestrial existence. The close of the Devonian is marked by one of the five major mass extinction events of the Phanerozoic. The trilobites, brachiopods and ammonites all suffered considerable attrition.

The first vertebrates to spend most of their life out of water were the amphibians, which comprise a separate class in the phylum Chordata. The amphibians evolved from fish in the late Devonian, the evolutionary transition between the two groups being clearly shown in the fossil record. Amphibians are semi-aquatic; their eggs must be laid in water and they are exclusively swimming organisms during the juvenile stage of their life cycle, only later developing legs for movement on land and lungs for breathing. Amphibians became prominent members of terrestrial and semi-aquatic environments at the beginning of the Carboniferous. They occupied a number of ecological roles, some were essentially herbivorous while others were obviously carnivorous, and some types grew very large compared to their modern relatives. The position of the amphibians as the dominant land tetrapod (four-footed) group, however, was cut short by the ascendancy of the reptiles.

Recent evidence suggests that the reptiles arose over 340 million years ago. Carboniferous reptiles were quite small, typically growing only to the size of present-day lizards. Reptiles have a number of features that equip them for a terrestrial existence. Particularly significant is the *amniotic* egg which protects the developing embryo from desiccation and provides it with some nourishment. So for the first time, vertebrate animals were independent of standing water for reproduction. In addition, reptiles are covered with dry scales which inhibit moisture loss. They have more efficient lungs than amphibians, a more effective dentition and generally superior mobility out of water. This suite of features gave the reptiles a considerable advantage over the amphibians and as they became more prominent in late Palaeozoic times, the ecological dominance of the amphibians waned.

In the late Carboniferous, around 300 mybp, fossils appear which are

Fig. 14.4 A collection of fossilized fish of Devonian age. Note their superficially 'modern' appearance.

essentially reptilian, but which also have some features associated with mammals. Known appropriately as mammal-like reptiles, these animals were the likely immediate ancestors of the mammals. The therapsids were a prominent group of quite advanced mammal-like reptiles in the Permian; some believe they were endothermic, i.e. able to regulate body temperatures (page 88) and thus less dependent on environmental temperature than reptiles. The variety of therapsid types found in Permian deposits indicates a major adaptive radiation and in late Permian times they may have been the dominant type of land vertebrate.

Permian extinctions

The close of the Permian, the final period of the Palaeozoic era, is marked by a mass extinction generally considered to be more catastrophic than any other extinction event before or since. These extinctions were confined principally to shallow marine environments where it is estimated that over ninety per cent of all species died out.

Several explanations have been advanced to account for this phenomenon, but particularly favoured are those which invoke plate movement as a major contributory factor. This is because by the close of the Palaeozoic, the continental land masses had converged to form just one 'supercontinent' (Fig. 13.2). Coalescence of formerly discrete land masses would inevitably have reduced the length of coastline. This, coupled with a regression of seas off the continental shelves which occurred in the late Permian, must have caused a considerable reduction in the areal extent of shallow marine environments. There is a tendency for species diversity to be positively related to the area available for colonization, so it follows that a decline in species richness should be a consequence of habitat reduction. Again, it is modern ecological theory that is used to account for changes in the biosphere in the distant past.

An alternative explanation is that a reduction in habitat diversity, rather than total area, led to the extinctions. Species unable to cope with the narrower range of environments would not have survived. Other theories involve climatic change and lowered salinity as contributory factors, but neither by itself seems to have been a potent enough force to trigger the extinction patterns revealed by the fossil record. Of course, if it is believed that mass extinctions occur at regular intervals, as a consequence of asteroid impacts (page 162), then continental movement is relegated from the primary cause of extinction to an exacerbating factor.

The Mesozoic era

Life in Mesozoic seas

The Permian mass extinction affected invertebrate groups differentially. Bivalve and gastropod molluscs suffered less attrition than ammonite molluscs (Fig. 14.5), although the latter recovered well to become important swimming predators in Mesozoic waters. The hexacorals, a group of corals still prominent in reef formation, appeared in the Triassic and largely replaced the types of coral characteristic of Palaeozoic seas. The early Triassic can be regarded essentially as an interval of recovery for marine invertebrates, some types undergoing rapid diversification following their decimation at the close of the Permian.

Swimming vertebrates were also prominent members of marine ecosystems during the Mesozoic. In Triassic times, fish were joined by marine reptiles (Fig. 14.6) which had evolved from terrestrial types. One group of marine reptiles, the nothosaurs, had paddle-like limbs and probably spent part of their lives out of water as seals do today. Their descendants, the carnivorous plesiosaurs, were more fully aquatic, some species exclusively so. Crocodilians were marine predators from early Jurassic times, having descended from principally terrestrial crocodiles that appeared in the Triassic.

Two types of unicellular alga, the dinoflagellates and the coccolithophores, were seemingly the dominant photosynthetic organisms in Mesozoic waters. Both of these groups are still very well represented in the phytoplankton.

Fig. 14.5 An ammonite, *Asteroceras*, from the Lower Jurassic, Humberside, eastern England. Ammonites were prominent marine predators in the Mesozoic but were extinct by the close of the era.

Fig. 14.6 Model of an ichthyosaur, a carnivorous marine reptile of Mesozoic times.

Mesozoic plants and the origin of the angiosperms

Land plants were seemingly not greatly affected by the events that caused the catastrophic losses of shallow-marine life at the close of the Permian. Seed ferns and, particularly, true ferns were common in Triassic floras, but the former group gradually declined in importance. For much of the Mesozoic, forests were dominated by flowerless seed plants, a group known collectively as gymnosperms. The word gymnosperm, of Greek origin, means 'naked-seed', and is used because the seeds do not develop within an *ovary*. Important Mesozoic gymnosperms were cycads, cycadeoids, ginkgos, and conifers, the last becoming progessively more prominent during the Jurassic and early Cretaceous. Although the cycadeoids are extinct, species of cycad, which have palm-like leaves remain, confined to tropical and subtropical locations.

None of the seed plants mentioned above produce true flowers. The flowering plants, the angiosperms, did not undergo their first major radiation until mid-Cretaceous times, around 100 million years ago. Angiosperms differ from gymnosperms such as conifers in several respects. The reproductive structures of angiosperms are flowers (Fig. 14.7), not cones. Whereas gymnosperm ovules, which develop into seeds after fertilization, lie unprotected on cone scales, angiosperm ovules are embedded within an ovary. Angiosperm pollen is received by the *stigma*, which is part of the female organ, not adjacent to the female gametophyte as in gymnosperms. If conditions are suitable, angiosperm pollen that alights on a compatible stigma develops a pollen tube that extends inside the style and carries the male gametes to the female gametophyte. (Some advanced gymnosperms also produce pollen tubes, but they are shorter than those of angiosperms.)

Unique to the flowering plants is the double fertilization that ensues between sperm and female nuclei. The pollen tube carries two sperm nuclei. One of these fertilizes the egg nucleus in the embryo sac, giving rise eventually to the seed embryo; another nucleus unites with a diploid nucleus (formed from the fusion of two haploid nuclei) in the female gametophyte. This second fertilization gives rise to the *endosperm*, a nutrient-storing tissue (obviously made up of triploid cells). Nutrients mobilized from the endosperm can support the development of the embryo and, in some types of plant, seed germination and early growth. Also, in contrast to the 'naked' seeds of gymnosperms, angiosperm seed develops within a fruit, which develops from the ovary wall. Depending on the nature of ovary wall development, fruits are categorized as either 'dry' or 'succulent'. The term fruit is also used informally to refer to any seed-containing structure, perhaps consisting of several true fruits or including other floral parts.

Several aspects of angiosperm origins remain controversial. The conventional view is that the angiosperms arose early in the Cretaceous, or possibly the late Jurassic, but a minority view advocates a much earlier origin. There is no consensus either concerning the immediate ancestors of the angiosperms and different types of seed plant have been suggested as candidates. Whether the first flowering plants were tree-like or shrub-like, and whether they evolved first in seasonally-dry or permanently-moist environments are two of the issues debated.

Flowers are not durable structures and claims made for the antiquity of flower fossils are not universally supported. Evidence for the evolutionary origin of the angiosperms is therefore largely based on other tissues such as leaves, stems, roots and, particularly, pollen. Fossilized

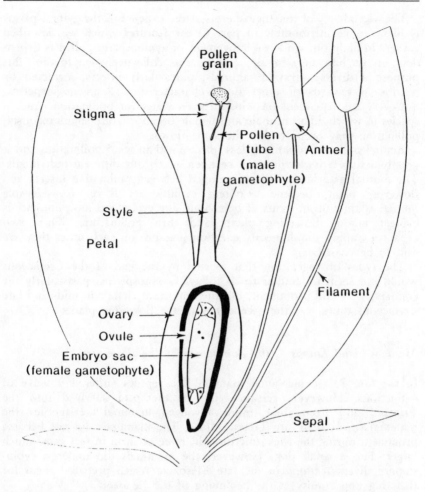

Fig. 14.7 Section through a 'complete' flower, i.e. one with sepals, petals, stamens (anther and filament) and one or more pistils (ovary, style and stigma). Pollen grains are produced within pollen sacs in the anthers. A pollen grain, landing on a compatible stigma, develops a pollen tube which extends inside the style and carries the male gametes to the female gametes. An ovary with one ovule is shown, but there may be several. The seed develops from the ovule, the fruit develops from the ovary.

specimens of these tissues are compared with those from known angiosperms in order to judge whether they belonged to flowering plants. A problem here is that such specimens are often poorly preserved, and anyway there is no certainty that the fossilized 'angiosperm-like' plant did actually bear true flowers. Notwithstanding these difficulties, there is no doubt that the angiosperms diversified rapidly during mid-Cretaceous times and had become the dominant group of land plant by the close of the period.

The ascendancy of the angiosperms at the expense of the gymnosperms is likely to be attributable in part to the features which we described earlier. In addition, while the movement of gymnosperm pollen is dependent on its buoyancy in air, angiosperms collectively employ for this purpose a diverse array of animals, particularly insects, attracted by nutritional rewards or other floral characteristics. Some angiosperms, however, are dependent on wind or even water for pollination. And in species in which adjacent male and female organs mature simultaneously, pollination may occur within the same flower.

Animal pollination is much less uncertain than wind pollination and it greatly assists cross fertilization between genetically different individuals. The mutual associations that developed between pollinating insects and flowering plants became a potent evolutionary force. Considerable mutual adaptation in terms of behaviour, life cycles and morphologies is evident among flowering plants and their pollinators. When two organisms apply simultaneous selection pressure on each other they are said to be *coevolving*.

The types of vegetation that existed by the end of the Cretaceous would not seem unfamiliar to us today. Gymnosperms, particularly the conifers, were still prominent, but the important change in mid- and late Cretaceous times was the ascendancy of the flowering plants.

Mesozoic land faunas – the dominance of the dinosaurs

In the late Palaeozoic some mammal-like reptiles suffered a wave of extinctions. However, certain types of therapsid survived into the Triassic, and from these more advanced mammal-like reptiles the mammals evolved in late Triassic times. The mammals did not become prominent during the Mesozoic though; none of them in fact grew much larger than a small dog. However, the mammals did undergo evolutionary diversifications in the late Mesozoic which prepared them for their big opportunity at the beginning of the Cenozoic.

The symbol of Mesozoic life is of course the dinosaur. The dinosaurs rose to prominence in the Triassic, dominated terrestrial environments throughout most of the Jurassic and Cretaceous periods, but then experienced a wave of extinctions towards the close of the Mesozoic, 65 million years ago. The dinosaurs descended from a group of reptiles called the thecodonts. Early dinosaur species were comparatively small, but by the late Triassic some species were typically six metres or more in length. In the early Jurassic the very large members of this group began to appear, some of which exceeded 30 metres in length. Some types of dinosaur are shown in Fig. 14.8.

The dinosaurs were a very diverse group of organisms and certainly not all species were as massive as their popular image sometimes suggests. Conventionally they are classified into two main groups based on hip (pelvic) structure. In one group, the ornithischians, the hip anatomy is not unlike a modern bird, whilst in the other group, the saurischians, the

Fig. 14.8 Models of some Jurassic dinosaurs. From left; *Stegosaurus*, *Apatosaurus*, *Allosaurus* and *Camptosaurus*.

hip structure resembles that of a lizard. In both groups there were species which moved essentially on two legs and species which were principally 'four-legged'. All bird-hipped dinosaurs were seemingly herbivorous, but the lizard-hipped group contained both herbivores and carnivores.

One facet of dinosaur biology to attract interest and speculation concerns their energy metabolism. As explained earlier (page 88), most types of organism are ectothermic, meaning that body temperature tends to equilibriate with the temperature of the environment. Mammals and birds in contrast are endothermic; they use a proportion of their food energy to maintain body temperature within a narrow range. A capacity to regulate body temperature permits activity over a wider range of environmental temperatures than is possible for ectotherms. It is argued that the dinosaurs could not have been so successful had their activity patterns been controlled by external temperature, particularly in competition with the small mammals with which they co-existed.

Details of bone anatomy have been used in support of the view that the dinosaurs were endothermic. In addition, the principles of food-chain energetics (page 104) have been invoked. The argument is based on evidence that the ratio of dinosaur predators to their prey in fossil assemblages is closer to that typical of communities in which mammals, not reptiles, are the dominant predators. The ratio of predator to prey production should be higher when reptiles are the predators because mammals require more food in order to maintain body temperatures. It is an interesting idea but we should be aware of the problems involved in determining the abundance of animal populations that existed over 65 million years ago. One problem in accepting the endothermic dinosaur theory is the enormous amount of food energy that would have been required to maintain body temperatures. Although not universally agreed, the dinosaurs are still usually classified as reptiles.

Whatever the truth about their energy metabolism, the dinosaurs were an extremely successful and well-adapted group. Fascination with their demise in the Cretaceous, which we will consider in the context of Cretaceous extinctions in general, tends to obscure this fact, but we should not forget they were the dominant group of land animal for nearly 150 million years.

The earliest flying vertebrates were reptiles, and they first appear in the fossil record during the late Triassic. Some from later in the Mesozoic had wingspans far in excess of any extant flying organism. Feathered birds, which comprise a separate class of vertebrate, probably originated in the late Jurassic. Their structure indicates a reptilian ancestry, but because their parts do not fossilize well, the early evolutionary history of birds is sketchy.

Late Cretaceous extinctions

The late Cretaceous is marked by a major mass extinction. This event is well-known because it accounted for the dinosaurs, but several other taxa, particularly of marine organisms, also experienced heavy losses. In the oceans, pelagic (open water) organisms suffered much greater losses than benthic (bottom dwelling) organisms, although for many groups there is uncertainty as to how suddenly extinction occurred. The marine reptiles and ammonites became extinct, and the diversity of planktonic organisms dropped markedly. The comparatively sudden loss of so many planktonic species suggests a sudden change in their physical environment, such as a cooling of oceanic water or a drop in salinity.

One theory is that an enormous asteroid struck the Earth, creating a dust cloud of sufficient thickness and duration to obscure sunlight, and hence prevent photosynthesis, and to reduce temperatures. Such an event, it is argued, would have drastically reduced the productive capacity of the biosphere and affected all organisms to a greater or lesser extent. There is evidence that such an impact occurred. Some sediments deposited at the end of the Cretaceous, in different geographical locations, contain an unusually high concentration of iridium, a chemical element that is extremely rare on Earth. Such unusually high concentrations of iridium have led some to suggest an extraterrestrial origin. But even if we accept this explanation for the iridium, it does not necessarily follow that the impact was responsible for the extinctions. Furthermore, not all types of organism were seriously affected in the late Cretaceous, and the extinctions that did occur were not necessarily simultaneous. All dinosaur species did not become extinct at the same time, and a few may possibly have survived into the early Cenozoic. So the debate about what caused the late Cretaceous extinctions continues. But whatever the cause, the consequences for the biosphere were enormous.

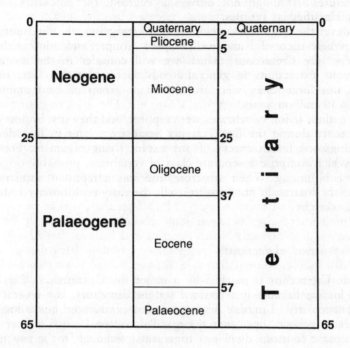

Fig. 14.9 Schemes in use for dividing the Cenozoic era into named intervals. Numbers are millions of years before present.

The Cenozoic era

Ascendancy of the mammals

The Cenozoic, from the Greek meaning 'new life', covers the most recent 65 million years. This era is very much shorter than either the Palaeozoic or the Mesozoic and its fossil record is more revealing. Different schemes, with different terminologies, are in regular use to divide the Cenozoic into intervals, which can be a little confusing. The most common are shown in Fig. 14.9.

The mammals, which had been inconspicuous members of Mesozoic faunas, underwent a spectacular adaptive radiation early in the Cenozoic to become the dominant type of land vertebrate. Whatever the reasons for their demise, the disappearance of the dinosaurs provided a situation conducive to mammalian diversification and expansion. Mammals are recognized by combinations of features, including glands on maternal parents from which milk is supplied to young offspring, a covering of body hair, a capacity to regulate body temperature and, particularly important in fossil remains, features of the jaw, dentition and ear. In the largest of the three mammalian groups, the placentals, the young are born alive at a fairly advanced stage of development. In contrast to the placental mammals, the marsupials release offspring into a pouch at a relatively early stage of development. Marsupials are the most conspicuous mammals in Australia and New Guinea and occur naturally elsewhere only in the Americas. The most familiar of Australia's diverse marsupials are the kangaroos and wallabies. The opossums of the Americas are also marsupials, but here the diversity of the marsupials has declined considerably during the Cenozoic. The least common mammalian group, the monotremes, are now represented only by the duck-billed platypus and echidna and are confined to Australia and New Guinea. Although monotremes possess several mammalian features they also lay eggs, as reptiles do.

By the beginning of the Cenozoic, the configuration of land masses (Fig. 13.2) was such that mammalian faunas on different continents evolved more or less independently. Much later there was to be considerable faunal interchange between some continents, particularly between the Americas and between North America and Eurasia, but nevertheless, that early period of regionally-independent evolution left a legacy which is still discernible. Mammalian size appears not to have increased very much during the first ten million years of the Cenozoic, but diversity greatly increased and many modern mammalian orders can be traced to this relatively brief interval. The rodents, an extremely important order of small mammals, date from the mid-Palaeogene.

The course of mammalian evolution was influenced by the progressive cooling and increased seasonality of climates that occurred from mid-Palaeogene times. The area of closed forest declined in response to these climatic changes while that of savanna-like and more open vegetation expanded. The grasses appeared in the Eocene epoch and eventually came to dominate the open landscapes that later developed. Grasses are well adapted to withstand grazing because vegetative growth occurs predominantly from around ground level. So grazing, unless severe and prolonged, tends to stimulate regrowth.

The expansion of open landscapes provided new evolutionary opportunities for the mammals, particularly the hoofed mammals (the *ungulates*) but also brought new challenges. For example, many grasses contain silica, a tough and abrasive mineral. Certain features of herbivore dentition seemingly evolved in response to a fibrous, largely grass diet, including tough enamel ridges to maintain an effective grinding surface, continuously growing teeth and high-crowned teeth to provide a large grinding area. Associated also with a fibrous diet is cud-chewing, i.e. the regurgitation of food from the stomach, where it is temporarily stored, to the mouth where it is processed further. Thus the animal spends less time than would otherwise be necessary in grazing on open grasslands, and digestion is completed when and where the animal will be less vulnerable to predation. The danger of predation has doubtless selected for speed of movement: in some types of grazing herbivore, the leg bones became progressively better adapted for faster movement during late Palaeogene and Neogene times. A capacity to utilize cellulose, the major chemical component of plant cell walls, is another characteristic feature of hoofed herbivores, particularly the ruminants. As we discussed in Chapter 8, cellulose can be utilized by such an animal because microbial residents of the digestive tract produce enzymes that break the bonds linking the component sugar units. Although mutualistic associations between animals and cellulose-digesting microbes did not appear first in hoofed herbivores, it is within this group of animals that such associations are best developed.

So in mid-Cenozoic times a suite of characteristics developed amongst hoofed, herbivorous animals enabling them to cope with a grass-rich, fibrous diet and an open environment. During mid-Palaeogene times the dominant ungulates were the odd-toed type (today, rhinoceros, horse and tapir), but in the later stages of the Palaeogene, the even-toed, or cloven-hoofed, ungulates (now for example sheep, cattle, camels, antelopes and pigs) assumed dominance, a position they have held ever since.

Mammalian evolution was not confined to land; the whales appeared, as descendants of land mammals, quite early in Palaeogene times and underwent an adaptive radiation in the Neogene. Other mammals which appeared in the Neogene, the seals, walruses and sealions, are pre-dominantly aquatic, although they reveal their terrestrial ancestry by coming ashore to breed. Mammals also took to the air in the early Palaeogene. Bats are the only flying mammals; they now have an almost world-wide distribution and they comprise one of the largest mammalian orders.

The late Cenozoic

The global trend towards cooler, drier and more seasonal climates was maintained during the Neogene and was accompanied by a progressive increase in open, grassland environments. Against this background the mammals evolved progressively more 'modern' characteristics. Neogene grasslands must have supported vast herds of grazing herbivores together with their predators; today we can catch a glimpse of such communities only on the remaining grasslands of eastern and southern Africa.

Global cooling became much more marked around 3 million years ago, heralding the onset of the interval commonly referred to as *the* Ice Age. In fact though, it is just the most recent of several ice ages that have occurred during the history of the planet. Ice ages are extended intervals of global cooling when snow and ice cover extensive areas for long periods of time. The general direction of ice movement is from high to lower latitudes, although mountain glaciers ahead of the main continental ice sheets also expand. Ice sheets have a profound effect on the land surface beneath and in the immediate vicinity giving rise to characteristic landscapes.

The most recent ice age occupies the interval known as the Pleistocene. Its commencement is usually set at about 2.0 mybp, although continental ice sheets may have started to form a million years earlier. At its maximum extent, ice covered about thirty per cent of the Earth's surface (compared with less than ten per cent now). However, the ice cover was not continuous throughout the Pleistocene; several minor retreats (*interstadial* phases) and some major retreats (*interglacial* phases) having occurred. The series of glacial advances and retreats provides the basis for subdividing the Pleistocene into named intervals. It is important to be aware that climatic changes during the Pleistocene were not confined to glaciated and adjacent areas but were global in extent. Outside the areas of immediate glacial influence, glacial advances were typically associated with cooling and often increased aridity, although some areas actually became wetter.

An accumulation of snow and ice during glacial phases is associated with a drop in sea level, thus permitting species migrations between previously separated land masses, as for example, between Alaska and Siberia, and between Australia and New Guinea. In the northern hemisphere, there was a trend for species to migrate more or less south and north in response to alternating cooling and warming, with consequent shifts in the composition of the biota in particular regions. However, we should not automatically conclude that whole communities shifted more or less intact. Rather, each species would have responded independently to climatic changes, so it is quite probable that species assemblages formed that are rather different from those present today.

The most recent 10,000 year interval comprises the Holocene, or Recent, epoch. The frequently-used term 'post-glacial' is misleading because we are living not in post-glacial times but in an interglacial. The significance of the 10,000 year date is that it marks the onset of rapid climatic amelioration in the northern hemisphere.

Late Pleistocene mammalian extinctions

In many parts of the world, large mammal faunas (*megafaunas*) suffered significant extinctions in late Pleistocene or early Holocene times. Of course, evolutionary 'dead-ends' are not uncommonly revealed by the fossil record and extinct species can sometimes be fitted into a phylogenetic scheme in which one replaces another. The late Pleistocene extinctions were different though, because during quite brief intervals a significant number of species of large animal became extinct.

The causes of these extinctions are a continual source of debate and speculation. It is tempting to regard climatic change as the principal cause, but we have then to explain, first why extinction was confined to large mammals and, second, why extinction on a similar scale did not accompany climatic changes earlier in the Pleistocene. Late Pleistocene extinctions involved animals that have close living relatives, but our familiarity and identification with these animals goes further because they were contemporaries of our own species, *Homo sapiens*. Inevitably then, we must consider to what extent human activity may have contributed to their demise.

Fig. 14.10   Woolly mammoth; one of a number of large mammals that became extinct in late Pleistocene or early Holocene times.

For North America in particular there is some support for the view that hunting by humans contributed significantly to the numerous mammalian extinctions that occurred at the very end of the Pleistocene. This view has been championed by Paul Martin of the University of Arizona, whose theory incorporates estimates of the size of contemporary human populations, and the possible impact they may have had through hunting. Exactly when peoples from north-eastern Asia first crossed the Bering land bridge during the late Pleistocene is debatable, but human activity was probably widespread over much of the North American continent by the time the first wave of extinctions occurred around 11,000 years ago. Within just a few millenia most of the larger mammals, including the ground sloth, camel, peccary, horse, mammoth, mastodon and the great plains cat became extinct in North America. In South America too, extinctions evidently followed human colonization. Large herbivores of South American origin disappeared together with many species of North American ancestry which had migrated south after the two continents had coalesced at the Panama Isthmus some three million years earlier.

Pleistocene extinctions also occurred in northern Europe and Asia; for the most part they preceded those in the Americas and were spread over a longer interval. Late Pleistocene inhabitants had included the woolly mammoth (Fig. 14.10), woolly rhinoceros, musk ox, giant deer, bear, bison, cave lion and cave hyena. During a few millenia, commencing about 25,000 years ago, many of these species became extinct, either completely or in Eurasia. Although these extinctions may have coincided with the emergence of hunting peoples in Eurasia, there is less enthusiasm for the hunting explanation than is the case for American extinctions, partly because it is not easy to explain why species likely to be favoured for eating, e.g. the red deer, survived while those less attractive from a dietary point of view, e.g. the woolly mammoth and woolly rhinoceros, became extinct.

The Australian megafauna, which included much larger marsupial mammals than at present, also experienced considerable attrition during the Pleistocene. This continent was first colonized by humans well over 40,000 years ago, although probably to a limited geographical extent. The major wave of extinctions seemingly occurred within this unique fauna around 30,000 years ago, and again took just a few millenia. (On the islands of New Zealand the case is rather different. Here the megafauna was dominated by large flightless birds and extinction is a much more recent affair, beginning about 1200 years ago. Again though, humans were responsible.)

The case of Africa is interesting, because it is in parts of this continent that present-day mammalian megafaunas are at their most diverse. Yet this is the continent with the longest history of human occupation, so why were African mammals not decimated also? In fact it seems that the African megafauna was more diverse during mid-Pleistocene times than is the case today, so perhaps here too human activities were important.

If humans were involved in late Pleistocene extinctions, by what means did they accomplish it? We should not overlook the fact that some of these animals were quite imposing, particularly bearing in mind the level of technology pitted against them. Another possible influence is habitat change, particularly through burning. Human activities may have contributed in more subtle ways, perhaps by altering predator-prey relationships or competitive balances between species. Clearly, there are no easy answers to the problem of late Pleistocene extinctions and different factors may have been responsible in different regions. Nevertheless, the fact that early human societies witnessed these extinctions cannot be disregarded, so explanations which do not consider human involvement are not entirely convincing.

The origins of the human species

Our own species, *Homo sapiens*, was introduced in the preceding section in the context of late-Pleistocene mammalian extinctions. Now we shall look at our own origins a little more systematically. Like so much that has been discussed in the historical part of this book, we are again in the midst of uncertainty and debate. The fossil evidence is simply not good enough to provide a detailed and unequivocal picture of human origins and ancestry. Opinions differ concerning the taxonomic treatment of the relevant fossils and their place in phylogenetic schemes. In fact, all the problems of relationships between taxa and taxonomic status introduced in Chapter 6 are much in evidence here. Different books and articles, especially if published over a range of dates are likely to show different schemes. It is important to be aware in this connection that the remains

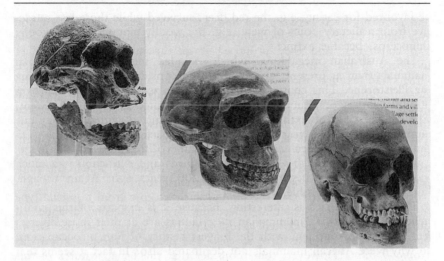

Fig. 14.11 Some hominid skull casts. From left: *Australopithecus africanus*, a gracile type of australopithecine; *Homo erectus*; *Homo sapiens sapiens*, the modern human type.

of our recent ancestors are typically just fragments of a skeleton or skull, so there is plenty of scope for different interpretations of their taxonomic status. Here we will attempt just a workable framework for human origins, including only those points on which there is considerable agreement.

Taxonomically, *Homo sapiens* subspecies *sapiens* is the only living representative of the family Hominidae. This family, whose members are called hominids (Fig. 14.11), includes extinct *Homo* species and species of the extinct genus *Australopithecus*. Hominid remains have been dated to about four million years ago. Hominids, together with the two ape families Pongidae (presently chimpanzees, gorilla, orang-utan) and Hylobatidae (gibbons) form a group called the Hominoidea. (Note the distinction between hominids and hominoids.)

The modern apes and the hominids evolved independently. However, it is not known when the lines leading to the hominids diverged from the other primates as the fossil evidence is poor for the critical interval 7.0–4.0 mybp. The fossil record for apes, or at least ape-like, animals extends back to around 24 mybp.

The earliest known hominids belong to the genus *Australopithecus*. Since the first australopithecine find, recognized as such by Raymond Dart in South Africa in 1924, controversy has surrounded this taxon, particularly its boundaries, taxonomic treatment of its members and its relationship to the genus *Homo*. Importantly, australopithecines were bipedal, whereas members of the ape line from which they diverged moved predominantly on four legs.

The oldest known type of australopithecine, now generally called *Australopithecus afarensis*, has been found in Ethiopia and Tanzania, and probably lived between 4.0 and 3.0 million years ago. This taxon was named in 1978 by American palaeoanthropologist Donald Johanson and colleagues, and they proposed it as the ancestor for all other members of the genus *Australopithecus* as well as the genus *Homo*. Others had favoured the view that the two genera evolved independently. Australopithecines probably lived approximately between 4.0 and 1.0 mybp and only in eastern and southern Africa. It is generally accepted that the genus *Australopithecus* should comprise a handful of species. These are designated as either 'gracile' or 'robust' forms. The robust form is characterized by a more exaggerated dentition, believed to be associated with a coarse plant diet. Gracile types, of which *Australopithecus afarensis* is one, preceded the robust form.

It is widely believed that the human lineage arose from a gracile type of australopithecine, and that before the australopithecines disappeared different hominids co-existed. Most authorities accept that the *Homo* lineage should comprise three species. The oldest of these is known as *Homo habilis*. (However, some consider that the facial features of this taxon are closer to the australopithecines than to the other two members of the genus *Homo*, and accordingly it should belong to the former genus.) Finds of this taxon during the early 1960's, in Tanzania's Olduvai Gorge, are associated particularly with the Leakey family who through their work in East Africa have contributed so much to knowledge about human origins. The epithet *habilis*, meaning 'able', was chosen because stone tools were found in the vicinity of some remains of this type. *Homo habilis* probably originated about 2.5 mybp and was fairly widely distributed over southern and eastern Africa.

There is seemingly some overlap between *Homo habilis*, which lived until 1.7 million years ago, and the larger and bigger-brained *Homo erectus*. Remains of *Homo erectus* (formerly often referred to as *Pithecanthropus*) have been found not only in Africa but in India, China, Java and, arguably, in Europe. *Homo erectus* used more advanced tools than *Homo habilis*, and from about 1.4 million years ago appears to have used fire as well. However, quite when fire was first used is a highly contentious issue. The most recent *Homo erectus* remains are about 0.4 million years old. There is then a gap of about 300,000 years before the earliest undisputed remains of our own species appear, again in East Africa. *Homo sapiens* is distinguished from *Homo erectus* by a number of features: on average brain size is greater, the lower jaw and mouth project less, and the forehead is higher and less browed. There are differences also in leg and pelvic anatomy.

The well-known Neanderthal man, named after the Neander River valley in Germany where remains were discovered in the middle of the nineteenth century, is now usually regarded as a subspecies of *Homo sapiens*, although there is a view that this type should be regarded as a separate species. The Neanderthals lived until about 35,000 years ago, but their fate and relationship to essentially modern *Homo sapiens*, known as Cro-Magnons, are still unresolved.

Much uncertainty surrounds the evolutionary relationships between the later hominids, and also their pattern and timing of diffusion. One view is that *Homo sapiens* evolved more than once from *Homo erectus* populations in separate geographical areas, but the majority view favours a single origin, probably in Africa, and subsequent diffusion throughout the Old World.

Postscript

The life-span of our own genus – roughly 2.5 million years – is just a tiny proportion of the 3.8 billion years or so that have elapsed since organisms first left traces of their existence on the planet. Translated into a time-scale that we can comprehend, this is equivalent to a little less than the final six hours of a calendar year.

It is hoped that in surveying important events in the history of life during these few chapters no hint has been given that the appearance of our own species represents an end point in any sense, or that evolution has a purpose and has finally achieved its goal. Certainly, we apply the terms 'mind' and 'culture' to no other living species, and no other shares our mental capacity, but this does not mean that we represent a logical outcome of all that went before us, nor does it mean that our origin was predestined. Expressed crudely and simply, it just happened that way. If the dinosaurs had not disappeared 65 million years ago we would not be here today.

Further reading

Eldredge, N. (1989) *Life Pulse – Episodes from the story of the fossil record.* London, Penguin Books.
(Excellent and very readable account of the history of life.)

Gould, S.J. (1980) *Ever Since Darwin.* (1983) *The Panda's Thumb.* (1984) *Hen's Teeth and Horse's Toes.* (1987) *The Flamingo's Smile.* London, Penguin Books.
(Superb anthologies of contributions to *Natural History Magazine.* Many articles on palaeobiological issues and much more besides.)

Halstead, L.B. (1982) *Hunting the Past.* London, Hamish Hamilton
(An authoratitive and well-illustrated introduction to palaeobiology.)

Lewin, R. (1989) *Human Evolution – An illustrated introduction,* 2nd edn. Oxford, Blackwell.
(Very clear account of hominid evolution – including cultural considerations – with a preliminary discussion of relevant theoretical issues.)

Margulis, L. and Sagan, D. (1987) *Microcosmos – Four billion years of evolution from our microbial ancestors.* London, Allen and Unwin.
(A popular treatment of the significance of evolutionary developments in the early biosphere.)

Stanley, S.M. (1989) *Earth and Life Through Time,* 2nd edn. New York. W.M. Freeman.
(Comprehensive, very fine account of the history of life in its geological context; lavishly illustrated. Highly recommended.)

van Andel, T.J. (1985) *New Views on an Old Planet – Continental drift and the history of the earth.* Cambridge, Cambridge University Press.
(Highly readable treatment of vital geological processes and the development of the biosphere.)

Glossary

Adaptation Any genetically determined characteristic (structural, functional or behavioural) that enhances survival and reproductive success in a particular situation.

Adaptive radiation Major evolutionary diversification giving rise to new taxa.

Aeon Largest formal division of geological time; the three aeons are the Archaean, the Proterozoic and the Phanerozoic.

Aerobic Of an environment in which free oxygen is present or a metabolic process involving free oxygen.

Algae Informal term covering a variety of photosynthetic unicellular or primitive multicellular organisms of (predominantly) aquatic environments.

Allele One of the alternative forms in which a gene occurs.

Alternation of generations Phenomenon whereby life cycle of plants and some algae comprises diploid (the sporophyte) and haploid (the gametophyte) phases.

Anaerobic Of an environment with no free oxygen or a metabolic process not requiring free oxygen.

Anaerobic respiration A type of energy metabolism which does not use oxygen but another chemical entity, e.g nitrate, as the terminal electron acceptor.

Angiosperm A flowering plant; a plant in which seeds develop from ovules within a closed ovary.

Anion A negatively-charged atom or group of atoms.

Artificial selection Deliberate selection of individuals for breeding.

Asexual reproduction Formation of a new individual organism without fusion of gametes.

Asteroid Rocky structure orbiting the sun; size range approximately 1.0–350 metres.

Asthenosphere Partially molten layer of the Earth's mantle on which the rigid plates of the less dense lithosphere 'float'.

Atom Smallest entity into which a chemical element can be divided and still retain properties of that element.

Atomic number The number of protons in an atomic nucleus; each element has a unique atomic number.

Autosome Any chromosome with the exception of a sex chromosome.

Autotrophic Of an organism, or type of metabolism, harnessing a non-biotic source of energy for organic matter synthesis from carbon dioxide.

Banded iron formation Iron-rich, laminated rock structure of Archaean and early Proterozoic times; laminated appearance is largely due to variations in reduced and oxidized iron content.

Basal metabolism All metabolic processes associated with normal functioning of an animal.

Benthic Of the zone, or the organisms, at the bed of a water body.

Benthos Community of organisms inhabiting the bed of a water body.

Catalyst A substance that enhances the rate of a chemical reaction but is not consumed during the process.

Cation A positively-charged atom or group of atoms.

Cell The basic living unit of which all organisms are constructed.

Cell wall Comparatively rigid layer surrounding cells of plants, certain protists and most bacteria.

Character Any feature of an organism that is used for taxonomic purposes.

Chemosynthesis An autotrophic mode of nutrition involving oxidation of simple inorganic entities to provide energy for organic matter synthesis from carbon dioxide.

Chlorophyll Green pigment which is the primary site of light energy absorption in photosynthesis.

Chloroplast Organelle in which chlorophyll is located within cells of photosynthetic eukaryotes.

Chromatid A longitudinal half of a chromosome formed by replication of DNA in early stage of nuclear division; the chromosome is then made up of two identical chromatids.

Chromosome Thread-like structure carrying genes; in eukaryotic cells located within the nucleus.

Codon A sequence of three nucleotides (a triplet) coding for a particular amino acid or a 'stop' instruction in amino acid sequencing.

Coevolution Evolution resulting from two taxa simultaneously exerting selection pressure on each other.

Community All, or a specified subset, of the populations of species inhabiting a prescribed area.

Compound A molecule consisting of atoms of two or more elements held together in a definite ratio.

Condensation reaction Chemical reaction involving the linking of two molecules with the elimination of water.

Constructive plate margin Boundary between two lithospheric plates at which new crustal material is extruded.

Continental drift Theory that the positions of continental land masses have changed through geological time; now accepted as fact.

Covalent bond A chemical bond resulting from the sharing of electrons between atoms or groups of atoms.

Cytoplasm The living part of a cell, excluding the nucleus in the case of eukaryotic cells.

Decomposition The breakdown and loss of dead organic matter.

Destructive plate margin Boundary between two lithospheric plates where one plate descends under the other and is destroyed.

Detritus The (little-altered) remains of living organisms.

Digestibility The proportion of a food stuff that traverses the wall of an animal's digestive tract.

Diploid Of a eukaryotic nucleus, cell or organism containing two sets of chromosomes; diploid number denoted by $2n$.

Dominant gene or allele Allele expressed phenotypically whether occurring in homozygous state or in heterozygous state with its recessive partner.

Ectotherm An organism without special features for regulating body temperature.

Electromagnetic radiation Radiation consisting of energy waves associated with electrical and magnetic fields; characteristics are related to wave frequency.

Electromagnetic spectrum The range of wavelengths and frequencies over which electromagnetic waves are transmitted.

Electron transport chain Term used to denote the passage of electrons between a series of special molecules to proton pumps where ATP is generated.

Element (chemical) A substance made up of atoms with the same number of protons.

Endergonic Of a chemical reaction requiring an input of energy from an external source.

Endosperm Nutrient storage tissue found in most angiosperm seeds.

Endotherm An organism with adaptations for regulating body temperature; notably mammals and birds.

Energy The capacity to do work.

Energy metabolism Metabolic processes involved in the capture, release and use of energy by an organism.

Energy substrates Chemical molecules which are broken down enzymatically to yield energy for metabolic purposes.

Enzymes Proteins that catalyze chemical reactions.

Epilimnion Relatively stable upper layer of low density water in a stratified water body.

Era Primary division of geological time in the Phanerozoic aeon; the three eras are the Palaeozoic, Mesozoic and Cenozoic.

Euphotic zone Zone of water in which light energy is adequate for photosynthesis.

Evolution Progressive change with time in the genetic composition of a population.

Exergonic Of a chemical reaction that liberates energy as it proceeds.

Fermentation An energy-yielding metabolic pathway not requiring oxygen and giving rise to a variety of by-products.

Fertilization Union of gametes with formation of diploid zygote.

Fitness A measure of the capacity of a genotype to leave its genes to future generations.

Flower The reproductive structure of an angiosperm.

Food chain and food web Diagrammatic representation of the movement of chemical energy and matter through a community of organisms.

Fossil Remains or other tangible evidence of an organism's existence in former times.

Fruit Mature ovary of a flowering plant (angiosperm); contains the seeds.

Gamete Sex cell; a special haploid cell or nucleus which unites with one of opposite sex to produce a (diploid) zygote.

Gametophyte (generation) Haploid, gamete-bearing, phase of a plant life cycle; may be dominant phase as in mosses or much reduced phase as in angiosperms.

Gene Sequence of DNA nucleotides which codes for a particular polypeptide or an RNA molecule. Genes provide information for structure and function in organisms and are passed on in reproduction.

Gene flow The movement of genes (strictly alleles) in and out of a population.

Gene frequency The relative abundance of genes (strictly alleles) in a population.

Gene pool The total complement of genes in a population.

Genetic code The universal genetic language made up of DNA triplets each of which codes for a particular amino acid or a stopping instruction in amino acid sequencing.

Genetic engineering The artificial transfer of DNA from one organism to another.

Genome The total genetic information in an organism's nucleus.

Genotype The particular combination of genes (strictly alleles) of an individual organism or cell.

Geological time-scale The formal division of time into named intervals; based primarily on the fossil record.

Gymnosperm A flowerless seed plant in which seeds lie unprotected, not within fruits.

Haploid Of a eukaryotic nucleus, cell or organism with only one set of chromosomes; haploid number denoted by *n*.

Heterotrophic Of an organism, or mode of nutrition, dependent on an external supply of preformed organic substances.

Heterozygous Having different alleles for a given trait on a pair of homologous chromosomes.

Homologous chromosomes The two chromosomes in a pair having identical gene sequences; diploid nuclei have the two homologues of each chromosome pair, haploid nuclei have only one of the homologues.

Homozygous Having identical alleles for a particular trait on a pair of homologous chromosomes.

Humus Chemically-complex and much-altered mixture of organic remains found in soil horizons and at soil surface.

Hybrid Offspring of two parents known to be genetically unlike in some respect; as in case of different genetic lines, varieties and species.

Hydrogen bond A weak kind of chemical bond involving hydrogen and (usually) two nitrogen or oxygen atoms; important in water and for holding large molecules such as DNA and proteins together.

Hydrolysis Chemical reaction involving the splitting of one molecule into two with the addition of hydrogen and hydroxyl groups from water to the products of the reaction.

Hydrothermal vent Site of hot-water spring on the ocean floor; associated with constructive plate margins.

Hypha (pl. hyphae) Filamentous structure which is part of a fungus.

Hypolimnion Lower layer of relatively high density water in a stratified water body.

Ice age Informal term referring to long interval of unusually extensive snow and ice cover.

Index fossil Fossil taxon used as reliable indication of age of rocks in which it is found.

Interglacial A comparatively long warmer phase of an ice age when considerable glacial retreat occurs.

Interstadial A comparatively brief warmer phase during an ice age.

Ion An atom, or group of atoms, carrying a net electrical charge on account of unequal numbers of protons and electrons.

Ionic bond Chemical bond formed by mutual attraction between oppositely-charged chemical entities.

Isotope One of two or more forms of the same element with a particular number of neutrons in the nucleus; isotopes of the same element thus have different mass numbers.

Joule Internationally recognized unit of work or energy.

Kinetic energy The energy possessed by a body by virtue of its motion.

Life cycle Genetically determined developmental pattern of an organism.

Lithosphere Rigid outer layer of the Earth; comprises crust and outer layer of mantle and fragmented into plates which 'float' on partially molten asthenosphere.

Maintenance diet The amount of food necessary to maintain an organism at constant weight.

Mantle That part of the Earth between the crust and the core.

Mass extinction Informal term for the disappearance of a significant proportion of taxa in a short interval of geological time.

Mass number Of an atomic nucleus; sum of number of protons and number of neutrons.

Meiosis A process involving two successive nuclear divisions during which the chromosome number in a nucleus is halved and there is opportunity for assortment of chromosomes and genetic recombination.

Metabolic pathway A sequence of enzyme-controlled biochemical reactions.

Metabolism The sum-total of chemical reactions occurring in a living cell, tissue or organism.

Metabolizable energy That part of ingested food energy available to an animal for metabolic purposes.

Micrometre One millionth of a metre (10^{-6} metre).

Mid-oceanic ridge Narrow submarine mountain chain – occasionally emergent – associated with constructive plate margins; notably Mid-Atlantic Ridge.

Mimicry Increased resemblance (usually form, colour or behaviour) of one type of organism to another type; usually a predator avoidance adaptation.

Mineralization Process by which elements held within organic matter are released in inorganic (mineral) form; occurs in soil and water during decomposition.

Mitochondrian (pl. mitochondria) Intracellular organelle which is the major site of aerobic respiration (oxidative metabolism); found in all but a few eukaryotic organisms.

Mitosis A somatic nuclear division that results in formation of two daughter nuclei with identical number of chromosomes and identical geneotype as the original; usually accompanied by cell division.

Molecule A stable chemical entity held together by chemical bonds and consisting of two or more atoms of the same or of different elements; a single molecule of a particular compound cannot be further divided and still retain the properties of that compound.

Moneran A member of the (prokaryotic) kingdom Monera; effectively a bacterium.

Mutation A change in the chromosomal DNA within a nucleus.

Mycorrhiza (pl. mycorrhizae) An association between a fungus and the root of a plant.

Nanometre One-billionth of a metre (10^{-9} metre).

Natural selection Differential reproduction among genotypes in a population in which there is variation between individuals and a potential for increase in number of individuals. Major mechanism for evolution.

Neutron Uncharged particle present in all atomic nuclei except common isotope of hydrogen.

Nucleus 1. Atomic: positively-charged central core of an atom, containing at least one proton and (except for hydrogen) at least one neutron; accounts for tiny fraction of an atom's volume but most of its mass. 2. Compartment in a eukaryotic cell containing the chromosomes.

Nutrients Elements particularly, but applied to all substances, essential for life processes.

Organelle A membrane-bound intracellular structure, e.g. nucleus, mitochondrion, which is the site of particular cellular functions; confined to eukaryotic cells.

Ovule Structure in seed plants comprising female gametophyte (including egg) and covering tissue; develops into the seed following fertilization.

Oxidation Chemical reaction involving the loss of electrons from a substance (which is thus oxidized). Metabolic oxidations often involve the addition of oxygen or loss of hydrogen.

Oxidative metabolism Energy-yielding process involving free oxygen as the terminal electron acceptor; essentially respiration.

Parasite Organism living on or in, and deriving its nutrition from, another living organism.

Peat The physically and chemically transformed remains of vegetation; its accumulation is due to the prolonged inhibition of decomposition processes.

Pelagic Of open water environments and organisms.

Period Primary division of time within geological eras in the Phanerozoic aeon.

Phenetic classification A classification of organisms based solely on resemblance and without deliberate reference to evolutionary considerations.

Phenotype The features of form, function and behaviour of an organism; an expression of interaction between genotype and environment.

Photosynthesis Biochemical and biophysical process involving the synthesis of energy-rich organic molecules from carbon dioxide and a hydrogen donor (usually water) using light as a source of energy.

Photosynthetically active radiation The range of wavelengths of the electromagnetic spectrum absorbed by chlorophyll and used in photosynthesis; approximately 400–700 nanometres.

Phylogenetic classification A classification of organisms based on evolutionary relationships.

Physiology The study of organism function.

Phytoplankton General term referring to microscopic photosynthetic organisms, mostly unicellular algae, in open water.

Plate Major unit of the lithosphere; plates are separated by different types of plate margins.

Plate margin Linear zone separating the plates which make up the Earth's lithosphere.

Pollen Male (micro)spores of seed plants; carries the male gametophyte and is transported by wind, animals or, sometimes, water.

Pollen tube The male gametophyte of seed plants; tube-like structure that develops as angiosperm pollen grain germinates on a receptive stigma; carries the male gametes to the site of fertilization in the female gametophyte.

Polymer A large molecule made up of much smaller molecules of similar type linked together to form a chain, as in polypeptide or polysaccharide.

Polyploid An organism having at least three times the haploid number of chromosomes; common in plants, rare in animals.

Population A group of organisms of the same species inhabiting a prescribed area.

Potential energy Energy in a potentially usable form.

Predation The consumption of living tissue by an organism; commonly used to imply capture of one animal by another animal.

Prey Any organism that is the food source of another organism; use sometimes confined to animals.

Primary production Organic matter of autotrophic organisms; principally photosynthetic organisms, but strictly chemosynthetic organisms also.

Primary productivity The rate at which organic matter is accumulated by autotrophic organisms.

Prokaryotic cell or organism One with neither duplicated chromosomes nor membrane-bound organelles; effectively a bacterium.

Protist A member of the eukaryotic kingdom Protista; most are microscopic unicellular organisms.

Radioisotope An unstable isotope that spontaneously disintegrates by the emission of particles and/or radiation.

Recessive gene or allele In heterozygous state, allele whose expression is masked by alternative, dominant allele and therefore does not contribute to phenotype.

Reduction A chemical reaction involving the addition of electrons or hydrogens to a substance (which is thus reduced).

Reproduction The formation of a new organism by either sexual or asexual means.

Respiration A series of biochemical reactions involving the release of energy from organic substrates for energy demanding activities; usually refers to oxidative metabolism but see anaerobic respiration.

Ribosome Subcellular structure comprising protein and RNA where polypeptides are produced in all types of organism.

Rumen Much enlarged stomach of many types of mammalian herbivore in which microbial populations anaerobically break down fibrous plant material and also synthesize amino acids from inorganic nitrogen sources.

Ruminant Any large herbivorous mammal possessing a rumen.

Sea-floor spreading The movement of crustal material either side of, and away from, a constructive plate margin on the ocean floor as new material is extruded; responsible for continental drift.

Secondary metabolite A chemical substance produced by an organism that is not directly involved in its own functional process but affects its relationship with other organisms.

Secondary production Organic matter of heterotrophic organisms.

Secondary productivity The rate of organic matter accumulation by (i.e. growth of) heterotrophic organisms.

Seed The fully mature ovule of gymnosperms and angiosperms; contains the embryo and (usually) a nutrient reserve.

Selection pressure Measure of the effectiveness of environment to change gene frequency in a population.

Sex chromosomes Sex-determining chromosomes; differ in type between males and females; possessed by most animals and some plants.

Sexual reproduction The production of a new individual by the union of two haploid gametes.

Solution A mixture formed by dissolving a substance (the solute) in a liquid.

Somatic cells All cells of an organism with the exception of the spores and gametes.

Spore Haploid cell of plants from which gametrophyte develops.

Sporophyte (generation) Diploid phase of a plant's life cycle on which spores are produced; alternates with gametophyte generation.

Stadial A cold phase during an ice age.

Stigma Part of female flower of angiosperms which is receptive to pollen grains.

Stomate (pl. stomata) Apparatus for gas exchange on plant leaf surface; comprises a pore (stoma) and surrounding guard cells; pore size varies according to water status of guard cells.

Stratified water body Water body in which distinct layering is evident; due to density differences induced by temperature and/or salinity.

Stromatolite Pillar-like limestone or, more rarely, siliceous structure up to a few metres in diameter with alternating organic-rich and organic-poor layers; former largely associated with cyanobacteria; prominent in Proterozoic aeon.

Subduction The descent of one lithospheric plate under another where two plates converge.

Systematics The study of all aspects of relationships between organisms.

Taxon (pl. taxa) A named group of organisms of any taxonomic rank.

Taxonomy That branch of systematics dealing with the classification of organisms.

Thermocline Relatively narrow layer of water in which temperature gradient is pronounced; separates the epilimnion from the hypolimnion.

Tissue A mass of cells which share structural and functional features.

Transcription First stage of protein synthesis process; involves 'reading' of section of chromosomal DNA and formation of complementary strand of messenger RNA which carries the information to the site of protein synthesis.

Translation Second stage of protein synthesis process; involves decoding information on messenger RNA molecule and linking of amino acids to form a polypeptide on the ribosomes.

Translocation Transport of soluble substances around a plant.

Transpiration The evaporative loss of water from the aerial parts (shoot) of a plant.

Transverse fault A type of geological fault which occurs where plates slide past each other without extrusion of new material or subduction.

Triplet A sequence of three nucleotide bases on a DNA or mRNA molecule.

Trophic level A clearly-defined stage in a food chain; e.g. primary producers – herbivores – carnivores.

Ultraviolet radiation Electromagnetic radiation in wavelength range 10–380 nanometres.

Ungulate General term for a large, hoofed herbivore.

Upwelling Upward movement of water from depth that occurs in some oceanic areas; important mechanism for nutrient transfer.

Vacuole Fluid-filled cavity characteristic of plant cells; often occupies much of cell's internal volume.

Vascular tissue Bundles of elongated cells which conduct water and inorganic nutrients and organic substances in solution, within plants.

Virus Entity comprising nucleic acid and (usually) protein coat; reproduction is confined to living cells in which a variety of responses, including many diseases, are induced.

Water potential The potential energy of water molecules; used as a measure of water status of plant tissue and soil; water potential is lowered with solute addition to pure waters.

Zooplankton General term for tiny animals and other non-photosynthetic organisms of open water which have little or no capacity for independent movement; includes juvenile stage of many organisms which later become free swimmers.

Zygote The diploid product of the union between male and female gametes; the first cell of a sexually produced individual.

Index